GW00976559

Bizarrchitecture

If only everything was as straightforward as it first seemed

Bill Swan

Published by New Generation Publishing in 2023

First Edition

ISBN: 978-1-80369-791-8

www.newgeneration-publishing.com

New Generation Publishing

Contents

BOGOF

Always look a gift horse in the mouth

I was going to write two books; one about the obscure career which I had pursued for over thirty years, and a separate one about the event which brought it to a premature end. Both are niche markets, and I was sceptical about the attraction to mass market of either. I wanted to write a book that readers would enjoy reading, and to share some personal experiences and insights which I felt might be of interest to others.

I have quite eclectic reading habits myself, often happening upon new, interesting fields serendipitously, and I felt that this might offer others an opportunity to discover a new, interesting but previously unknown, field.

So, what you get is a (hopefully fresh) insight into a field that you're interested in, and a free insight into a field that you didn't know you were interested in until you read this.

There is, of course, an outside chance that you are interested in both, in which case you really got lucky. If not, maybe you know someone who has an interest in the other part, and would like to go halves with you? Either way, it's all for the price of one book. Buy One, Get One Free.

Mythology

'There are two rules for success: 1) Never tell everything you know.' (Roger H. Lincoln)

For years there was a faint suspicion in our family that I had a secret penchant for Martini (shaken not shtirred) or would be revealed as some Harry Tasker character (played by Arnold Schwarzenegger) from the film True Lies. The flimsy foundation was that I studied Russian at the progressive Comprehensive school that I attended in the 1960's. It didn't seem to matter that I had failed the 'O' level, and have had 50 years since to forget. Oh, and the fact that I have signed the Official Secrets Act twice, even though, like all UK citizens, I had always been subject to it.

But my imagined role was seemingly confirmed when, in 2003, I joined a UK policing QUANGO[1]. Amongst the kit that I was issued with was a then state-of-the art PDA[2] which, my Son soon discovered, had previously been issued to someone who worked for MI5, and not all data had been scrubbed from it.

I then spent a decade doing a job that I couldn't describe well, which involved plenty of travel, including overseas and, towards the end of my career, I spent a year working away from home, in Swindon, which involved travelling there early on Monday mornings, staying locally during the

[1] Quasi Autonomous Non-Governmental Organisation - an organisation to which a government has devolved power, but which is still partly controlled and/or financed by government bodies – in this case PITO (the Police Information Technology Organisation) controlled by the Home Office

[2] Personal Digital Assistant. Specifically a Palm 'Tungsten'

week, and returning home for weekends. Swindon is, apparently, relatively close to Cheltenham, the home of GCHQ.

So, that settled it – I was a spy. If only life were that simple.

Interesting times

'A man who views the world the same at 50 as he did at 20 has wasted 30 years of his life.' (Muhammad Ali)

'May you live in interesting times[3]' – often incorrectly attributed as an ancient Chinese proverb - is said to be a curse, but I believe that I've lived through interesting times, and feel lucky to have done so. It would have been enough to avoid involvement in a major war, but my generation has been spoilt by all sorts of technological wizardry, including the emergence of affordable computing, mobile telephony, GPS navigation, and the World Wide Web. In fact most of what constitute 'the grid' off of which my generation grew up.

Then there have also been scientific and social revolutions such as the Moon landing and space flight, the virtual elimination of smoking, overcoming of killer diseases such as Polio, Measles, Smallpox, Rubella and AIDS, birth control, Social Media, 24hr, 7 day/week trading, home central heating, double-glazing, dishwashers, fitted carpets and toilet paper which isn't better suited to drawing on.

Meanwhile there has been a decline in democracy as social media and a politically partisan press have promoted "alternative truths" and a few major corporations have become more influential than many governments[4], a rise in

[3] First recorded use was by Sir Austen Chamberlain – brother of the British Prime Minister, in 1936

[4] These days, private individuals are as likely as nation states to be launching manned space missions.

"simplism"[5], and a vastly widening gap between rich and poor. The old adage 'poor people pay taxes, rich people pay accountants, and the super-rich pay politicians' has increasingly become reality.

During the first half of my career I was lucky enough to work in a wide variety of industries, for various employer types and sizes, and to have seen plenty of interesting changes at first hand.

Major job segues can offer tremendous opportunities to bring a fresh perspective to a role, but they also present huge challenges in gaining sufficient understanding of the business to earn the respect of colleagues, and to make yourself practically useful as soon as possible.

It would have been great if there had been ready points of reference, but the norm is usually at best a variety of dispersed and diverse resources and the contents of a few key experienced people's brains. Often those people were the least accessible.

I began to yearn for a 'bible' for each business – a blueprint which described its purpose, what it did and for whom, the rules which governed it, the organisation and responsibilities, the things that it consumed and produced and so on.

And the more of this stuff that is simply carried in the heads of people the greater the risk to the business. Most organisations will say that their most important assets are their people, ignoring the simple reality that 'their people' are not theirs – not corporate assets. They can and do walk when a better opportunity arises. Or they get lucky on the lottery, or unlucky on the road. Even a notice period

[5] the oversimplification of an issue

cannot be guaranteed – once somebody has decided to leave, they, and their knowledge and experience must be regarded as lost to the business. And, in the meantime, people know that the more they keep in their head, the more indispensable they are, so can be reluctant to give it up, or at least write it down.

Anyway, following are some tales from my own personal working life which demonstrated to me that experience is something which you don't get until just after you needed it[6], and which whetted my appetite for that elusive blueprint.

[6] Steven Wright

There are two types of people in the world - those who are able to extrapolate information from incomplete data

When I joined NatWest Bank, more than a dozen years after its formation through the merger of Westminster and National Provincial Banks, the rationalisation of its 3,600 branch network was still far from complete. Plenty of towns still had two (and sometimes more) NatWest branches. Sometimes because both had been important branches for the 'constituent' banks, sometimes because neither branch had been large enough to accommodate the combined business. Rationalsation of branches was an enormous, necessary exercise but another even bigger one was to come, even before the 'Big Bang' deregulation of 1986, and the reckless rush into new business areas (which was to be the eventual downfall of the bank).

It was called 'functional rationalisation', and was already well under way when I became involved, through banking support functions. I played only a minor part, but this was probably a seminal moment in my career - one which caused me to decide the direction I wanted it to take.

Historically, bank branches had been largely autonomous and self-sufficient, or associated to larger local branches which were. Most key functions could be undertaken locally, and they were. But there were efficiencies to be had.

The bank had grown organically for 400 years but, given the choice, would it start like that in 1970/80? With the business model, technology and customer demands of that world, how might it be designed? What activities needed to be regarded as 'core'? And so on.

I'm sure that somewhere, somebody was looking at how many Private Customer, Commercial Customer and Foreign Business cashiers were needed in each geographic area, but my interest was in support services.

For each 'function' or 'service' (for examples cheque book production, bullion services, central computing) what was the

- Projection of required capacity and geographical demand?
- Required hours of operation/expected service demand?
- Ideal location (need for physical proximity to other facilities, support infrastructure, costs, staff availability, other risk factors
- Required resilience?
- and so on

Each function (or service) represented a unique challenge.

Although I didn't fully appreciate it at the time, I had fallen, with great timing, into a field which both fascinated me and which I seemed to be good at.

During the next few years, against the backdrop of the dawn of desktop computing, then the birth of the worldwide web, I noticed a few examples of business issues - often masquerading as something else – where a structured approach to this kind of thing could be useful.

'An idiot with a plan can beat a genius without a plan' (Warren Buffet)

Armed with a "blueprint" of a business, a comparison between the current arrangement and ideal design for each function could be made, and necessary changes considered, prioritised, planned and executed.

In NatWest we ended up with, I think, five or seven newly constructed bullion centres, designed for optimal efficiency, distributed geographically to minimise distances for deliveries (but without offering too easy escape routes), and with good local emergency support services access. On the grounds that sometimes it pays NOT to advertise they were, thoughtfully, re-named 'Metal and Paper Packaging and Distribution Centres (MPPDC).

Computer centres were also 'dark sites' and, because of the increasing business dependence upon them, one of the key considerations was the avoidance of risk factors such as power lines (and magnetic fields), flood risk zones and aircraft flight paths – especially military. The last of these was vindicated when a Boeing 737 crashed at Kegworth in 1989, narrowly missing one of our vital centres, which had already been earmarked for replacement.

A key determinant for the location of a call centre was the accent of call-handlers. A Geordie dialect was found to be most trusted[7], so a new call centre was located in Newcastle.

[7] At least that was the story I was told. Interestingly, our Group Chief Executive at the time was a Geordie

'You are what you do, not what you say you'll do' (C.G.Jung)

The department responsible for the NatWest car fleet managed, through a series of clever strategies and policies, to run the operation at a profit. They allowed only a limited range of cars and specifications from which to choose - only those made in Britain, and only in metallic colours, for examples. They then negotiated bulk contracts direct with three or four manufacturers. Company cars were changed after just six months, meaning that they were always within manufacturers' warranty and rarely required any servicing, or even new tyres, brake linings or other 'consumables'.

We bought a LOT of cars (more, I was told, than Hertz) and a call-off schedule attractive to suppliers was agreed, so we bought at VERY competitive prices. The final step was feeding the little-used cars back into the market through a network of specialist outlets.

The team had even more ambitious plans. Reintroducing the used vehicle to the market represented a major challenge, but it also offered a significant opportunity for a new revenue stream, which would complement banking services offered to commercial customers. With a ready and predicable supply of 6 month-old vehicles we could offer packaged fleet services to customers, including the financing, insuring and maintenance (using the expertise that we already had – the Bank could become just another, albeit very large, customer). We already had, within the Group, Lex Vehicle Leasing, to manage the financing of customer vehicle fleets.

But this was a contentious proposal. The manner of operation alienated car dealerships, many of whom took their business elsewhere, together with access to their own customer purchase financing. It was a tough commercial trade-off and, in the event, the group didn't launch NatWest Vehicle Solutions until 1997 – after Lloyds, Barclays and HBOS (Midland) had launched similar offerings.

'Simplicity does not precede complexity, but follows it' (Alan Perlis)

In 2006, whilst leading the Conservative party in opposition, David Cameron described the police as "Britain's last great unreformed public service". I don't know about Britain's other 'great public services', but he was right about the police. Policing in England and Wales[8] is undertaken on a regional basis by 43 local forces (and the national British Transport Police), each a local fiefdom. Very few services were consolidated regionally or nationally - their own 'functional rationalisation' (as undertaken at NatWest), although there was plenty of opportunity. Cameron became Prime Minister following the Conservative Party's victory in 2010, but I haven't noticed much rationalisation since.

When I was working in the Home Office I was asked, in 2011, to work with a couple of colleagues to look into firearms licensing. Because only about 20% of the cost of managing the licencing of firearms was recovered through fees charged for licences it was seen as the general public 'subsidising toffs' sports', which was politically embarrassing, and a drain on police resources, so a target for savings.

Firearms control in the UK is among the toughest in the world, and the classification and licensing of weapons is complex. The British Police service is currently responsible

[8] Policing of Northern Ireland has long been carried out by a single force – the Royal Ulster Constabulary being replaced, following the Good Friday Agreement and the subsequent Patten Report, by the Police Service of Northern Ireland. At this time of our report the process of combining Scotland's 8 forces into a single service was well underway (Police Scotland, from 2013).

for managing the licencing of shotguns and firearms, and they do this on a regional basis, with each Chief Constable responsible for his or her own patch. Each has its own, independent, Firearms Licensing Unit (FLU).

The firearms landscape changed significantly with new Firearms Acts outlawing the private ownership of most handguns within the United Kingdom in 1997, as a result of the Dunblane school massacre of 1996[9]. 162,000 pistols and 700 tons of ammunition and related equipment were handed in, and the number of firearms certificates fell by over 24,000 by 2002.

There have been no shootings at UK schools since[10].

The issuing of firearms Certificates is based on firearm categorisation which reflects concealability, and which follows a rigorous vetting process. That process involving positive verification of applicant's identity, two referees of verifiable good character who have known the applicant for at least two years (and who may themselves be interviewed and/or investigated as part of the certification), approval of the application by the applicant's own family doctor, an inspection of the premises and cabinet where firearms will be kept (firearms/shotguns must be stored in a safe bolted to the floor or wall) and a face-to-face interview by a Firearms Enquiry Officer (FEO) also known as a Firearms Liaison Officer (FLO). A thorough background check of the applicant is then made by Special Branch on behalf of the FLU. A Firearms Certificate will only be issued when all

[9] On Wednesday 13 March 1996 a man armed with four legally-held handguns entered Dunblane Primary School and shot sixteen pupils and one teacher dead, and injured fifteen others, before killing himself.

[10] For comparison, in the USA there have been about 300 school shootings resulting in over 300 deaths since the Dunblane massacre.

these stages have been satisfactorily completed, and must be renewed every 5 years.

Justification must be provided to the police for each firearm and the police must be satisfied that the applicant has 'good reason' to own each one. Firearms are listed individually on the certificate by type, calibre, and serial number. A shotgun certificate similarly lists type, calibre and serial number, but the certificate permits possession of as many shotguns as can be safely accommodated.

There are, in the UK[11] today, about 1.4m shotguns held on almost 600,000 Certificates and about 600,000 firearms held on 150,000 Certificates.

Licenses are recorded on a system called NFLMS[12] which, in turn, is linked to the PNC[13].

Whilst looking at the activities covered, the methods, interactions and products, I was struck by the overall nature and culture of the units. They were staffed by two different types, using different skillsets to perform very different activities. The administration team maintained the records, processed new application documentation, ran the diarising of expiries and renewals, and scheduling of inspection visits and other work. The team of warranted police officers[14] tended to be comprised of policemen, nearing retirement age. They were highly experienced, knew their patch, and had an excellent network of contacts.

---

[11] For comparison, there are about 270 million guns in the USA.

[12] National Firearms License Management System

[13] Police National Computer

[14] Police officers in the UK swear an oath of allegiance to the Crown, are servants of the Crown and, in return, are granted a warrant. Each sworn constable is an independent legal official and each police officer has personal liability for their actions or inaction. Their warrant gives them additional legal powers of arrest and control of the public.

They vetted License applications and made site visits to check on weapon storage and site security.

The atmosphere in the ten or a dozen of the FLUs which we looked at appeared relatively relaxed, compared to operational policing, but the teams knew that they were fulfilling a vital police function, helping to ensure that firearms weren't in or likely to fall into the wrong hands, and making sure that firearms records were reliable, so that their operational colleagues could be made aware, at the time of operational deployment briefing, of the likelihood of firearms being present.

But the functional differences between activities were stark, and replicated 43 times. The administrators were not trained, warranted and experienced officers and the policemen were not employed for their administrative prowess. Those differences mattered most in the rationalisation options included in our report, which made clear that whilst the professional judgement aspects associated with the Warranted Officers were clearly core policing activities, the administrative functions were not and could be considered for centralisation, or even outsourcing[15].

In the event, our recommendations – including rationalising (centralising) the administrative functions and increasing firearms licence fees were not implemented – the latter presumably a political decision.

There remain rare cases where the system fails. Whilst writing this book – in August 2021 – a 22 year-old Plymouth man shot dead his mother, four others then

---

[15] There are, of course, already mature public licensing functions - including of vehicles and drivers.

himself. He had, apparently, had his firearm and license taken from him the previous December following an accusation of assault, but they were both returned to him by Devon and Cornwall Police service in July. This was despite 'Davison [having] posted hate-filled online rants about single mothers and about his own mother in particular'[16].

As I write, The Independent Office for Police Conduct (IOPC) is investigating the case.

---

[16] BBC: https://www.bbc.co.uk/news/uk-england-devon-58209726

'What gets measured gets done'.

I once attended a meeting of police Chief Constables whilst they were discussing what form a survey of public satisfaction and concerns should take. It was the practice of local forces to collaborate, through ACPO[17], by allocating shared challenges to particular forces or groups of forces to develop 'solutions' that could be shared by all. This was such a sub-group. The survey was an annual event, and their starting point was, logically enough, the previous year's format. The discussion centred around what had gone well and not so well the previous year, from which it was evident that the response rate had been disappointing, and the results not particularly enlightening.

It seemed to me that nobody in the room had much confidence in the whole exercise, and hence not much enthusiasm for what appeared to be a "box-ticking" exercise. Was it really going to form a sound basis for future policy-making?[18]

Firstly, only those with firm views tend to respond – polarised at opposite ends of the spectrum from love to hate.

Secondly, people frequently lie, and for a variety of reasons, some of which they're not even aware of (in which case I suppose it's not really 'lying'). Often they have an axe to grind, an agenda to pursue, a score to settle, a point to

---

[17] The Association of Chief Police Officers (a Private Limited Company), which provided a forum for Chief Police Officers to share ideas and coordinate operational responses. It was replaced in 2015 by a new body, the National Police Chiefs' Council.

[18] I believe that this task now falls to police and crime commissioners (PCCs) elected (4-yearly) in England and Wales and responsible for securing efficient and effective policing of a police area. The first incumbents were elected on 15 November 2012

make, a world-view to reinforce, and this is their only opportunity to do so.

Surveys are, in my view, the lazy way. Better to think about what you want to know, and then look for the best indicators. They are often already there - Like lichens on trees, which reflect the air quality of an area. I remember NatWest Bank's Chief Engineer saying to me once – 'if your lifts never break down, then you're paying too much for lift maintenance'[19].

In the case of policing, for examples, instead of asking women how safe they feel walking alone in an area during hours of darkness, why not just count how many are actually doing it? What proportion of local crime is even reported to the police? How do home and vehicle insurance rates in the area compare with national averages?[20]. You might need to think laterally. A favourite maxim of mine is ***'The best way to clean your fingernails is to wash your hair'***.

And beware of unintended consequences, or Perverse Incentives, as they are more formally known. Actions result in changes, which usually have multiple outcomes. Whilst the intended outcome might occur, a number of unexpected outcomes will often occur – some unwanted, and a few which outweigh any beneficial ones. It was the economist Horst Siebert who coined the term 'cobra effect' based on an initiative reputedly pursued by the British government, when concerned about the number of venomous cobras in Delhi. Apparently, a bounty was offered for every dead cobra.

---

[19] He (and, to a lesser extent, I) was responsible for the maintenance of the lifts in what was, at the time, the tallest building in Europe – the NatWest Tower

[20] Insurance companies already have a vested interest in knowing that kind of information

Initially, this was a successful strategy, and large numbers of snakes were killed for the reward. But enterprising people began to breed cobras for the income. When the government became aware, the reward programme was scrapped, and breeders set their cobras free, resulting in an increase in the wild population.

Mark Twain made reference, in his autobiography, to the phenomenon ***'Mrs. Clemens conceived the idea of paying George a bounty on all the flies he might kill. The children saw an opportunity here for the acquisition of sudden wealth. ... Any Government could have told her that the best way to increase wolves in America, rabbits in Australia, and snakes in India, is to pay a bounty on their scalps. Then every patriot goes to raising them'.***

And of course we're all familiar with the song 'I know an old lady who swallowed a fly'[21] (probably, like me, the Burl Ives version), and the ultimate consequences.

Another example, currently being seen in the UK, seems to be 'fly-tipping'. Local public services are reliant for their funding on taxes collected at a national level and, in order to meet national 'austerity' targets, that funding has been progressively reduced over the last decade. This has had various implications, as local authorities have desperately tried to balance their books by cutting back local services, or charging for them. One example is waste collection, where increasingly the disposal of waste at 'amenity tips'/'recycling centres' is often, now, charged for. It's an idea that looks appealing, initially, but is leading to increasing 'fly-tipping' to avoid the new additional cost

---

[21] There are many variations of phrasing in the lyrics, but The definitive version was written by Rose Bonne (lyrics) and Canadian/English folk artist Alan Mills and copyrighted in 1952.

(although I notice that our neighbouring County claims ***'We understand that making changes to our services raises concerns about the potential increase of fly-tipping. However, there is no clear evidence from other councils who charge for waste disposal that shows there is a link between charging and fly-tipping')***.

Shortly after I wrote the last paragraph I was faced with a perfect example, when I took my dog for an evening walk at a local beauty spot. The Local Authority currently charges £4 per sack to leave this sort of waste at an Amenity Recycling Centre (ARC), so this little lot (about a car load) saved someone about £30 and a drive to the nearest ARC.

When I reported it I was told that arrangements would be made with the Authority's contractor to attend to it. I don't know how much the LA "make" from the ARC charges, nor whether an analysis of the overall costs and benefits has been undertaken, but the cost of clearing up this rubbish (by a private contractor) must be significantly higher than dealing with the same material which is brought by the public to its recycling centres. And in the meantime, of course, we are treated to scenes like this.

Similarly, increased car parking charges can appear to offer an easy source of additional revenue for local authorities, but can lead to increases in illegal (and often dangerous or anti-social) parking, and unwelcome changes in visitors' and shoppers' habits. Some people may be dissuaded from visiting, or will change their shopping venue. Ultimately, towns can be "killed" in pursuit of a few extra shillings.

'To Err is human; to really foul things up Requires a computer' (Agatha Christie[22])

At NatWest we discovered that one of our engineers had, in his spare time, and at home, built a database[23] of all electrical devices – down to every lightbulb - in all our key buildings (London, at least). This 'skunkwork'[24] database had become business-critical, being used for total electrical load change evaluation and forward planning on, for example, all lighting scheme changes or the introduction of new equipment. The database was hosted on a desktop PC (appropriately one of those new bits of equipment which were springing up at the time) and supported by one man, the developer, who was a member of the Engineering team.

DataEase told us that it was the largest implementation of their product at that time. Kudos to the hobbyist whose baby it was, but it was a demonstration of the anarchy which distributed computing had become at that time. I'm not an engineer, but I can see the value of such a tool, especially in particular businesses, such as hotels, faced with uneven loads placed on electrical and heating supplies at particular times, as hundreds of guests simultaneously rise, take a shower and turn on their kettle for a quick cuppa whilst catching up on TV news headlines. It's one reason why hotel room kettles are such puny, 600-800watt affairs. Imagine the collective load of hundreds of standard 2-3Kw kettles being turned on at the same time. But is the

---

[22] Though sometimes misattributed to Paul Ehrlich

[23] Using DataEase – a relatively new RDBMS which enabled non-programmers to rapidly develop useful software applications.

[24] Everett Rogers, an eminent American communication theorist and sociologist, defined skunkworks as an "enriched environment that is intended to help a small group of individuals design a new idea by escaping routine organizational procedures"

same true in a fairly standard office environment during office hours?

If this tool really was "business critical" then it presented another set of problems which needed fixing. We would have to determine what specific needs it was intended to meet, how would data be sourced, verified, updated, what quality standards were required, backup and contingency arrangements, who would maintain and support it, what other systems did it need to interface/connect with, what processes needed to be introduced or modified, and so on.

'It is strange how new and unexpected conditions bring out unguessed ability to meet them.' (Edgar Rice Burroughs, The Warlord of Mars)

On Saturday, 24th April 1993 I was working, with a few members of my team, in our (NatWest) office in Kings Cross, north London. At about half past 10 the IRA detonated a 1 tonne ANFO[25] bomb in a stolen tipper truck in Bishopsgate, in the City of London. Although we were not much more than two miles away we didn't hear the blast, but it didn't take long for the 'phones to start ringing.

With the City cordoned-off by the emergency services, the Bank's executives decided to make our offices their base from which to work - to assess the business impact and plan immediate contingency arrangements and recovery. We were outside the damage zone, but relatively close, and had plan chests full of A0[26] size drawings and many details of all Bank buildings in London. We had, at that time, only a few desktop PCs, principally for word-processing and not yet any network-accessible contingency plan, nor even much in the way of operational data. This was a Premises department office, whose operational concern was for the buildings and services.

The Bank's executives arrived, in their weekend attire – including some straight from the golf course – and set about addressing the challenges. The first were trivial, but unforeseen, logistical problems, such as not having a key, and

---

[25] Ammonium Nitrate/Fuel Oil (94% Ammonium Nitrate) - the IRA's bomb of choice at the time. Although it had a devastating effect in the confined area, the world was reminded of its explosive potential in 2020 when 2,750 tonnes of Ammonium Nitrate exploded in a harbour warehouse in Beirut.

[26] The ISO 216 standard for paper sizes. A0 = 1189 x 841 mm

hence access, to the car park. With a long day's work ahead, the lack of on-site catering might have been a problem had it not been for an abundance of local fast-food outlets and the initiative of a resourceful member of the team, who took himself off and returned laden with bags full of fish and chips, bought on his personal credit card.

More telling was our discovery, having realised very early on that the key asset needed was the operational knowledge that existed most readily in key individuals' heads, was that those individuals needed to be called in, but that nobody had a list of their home 'phone numbers (mobile 'phones were still rare), and that the office switchboard had Directory Enquiries "locked out" at the weekend! The whole exercise was dependant on one of us going, armed with a pile of small change, to a public telephone box across the road. Only when we had managed to contact those key individuals and they had managed to make their way in, and more detailed damage assessments received could planning for opening for business on Monday morning be completed. That it was achieved is a credit to the team that day.

The explosion killed one person[27], injured over 40 people, and caused between £350m and £1bn worth of damage (depending on source). Police had received a coded warning, but were still evacuating the area at the time of the explosion. NatWest Tower was badly affected, although HSBC's building on the junction of London Wall and Bishopsgate was directly between it and the bomb, and took the brunt. Nevertheless, the damage to the NatWest Tower was substantial. A new Tannoy system had been installed in the building only a week or so previously and there were only about 50 or fewer staff working on that day. But the Police hadn't yet contacted the

[27] Ed Henty, photographer

building security staff, and it took the initiative of one - when a courier arrived saying that access had been difficult because the area was being cordoned off - to issue an instruction to all staff to immediately take refuge in the basement (there were 3 levels), which is where they were when the bomb was detonated. The initiative of that person undoubtedly saved lives. I went up the Tower the following week, and witnessed the devastation. Shards of window glass were embedded in the central concrete core, and work areas were a mess. I was harnessed to the core so that I could witness the fenestration damage up close, and see the blast site, and it was clear that casualties would have been horrific had the offices been occupied as usual. Strangely, the blast had travelled up the lift shafts and jammed the lift doors in their frames, so the evacuation to the basement had been timely.

A less predictable casualty of the blast was Drapers Gardens, on Throgmorton Avenue, where NatWest had located many of its Head Office functions. The fine, 1967, green glass 30-storey building[28] sustained damage from blast waves "funnelled" through back streets, which affected key support offices.

Future contingency planning focussed on the host of new business-critical technology, and was based on 'minimal' or 'essential' operational requirements which called into question why we had ever needed more, but that's a different story.

---

[28] Coincidentally, like NatWest Tower, designed by Richard Seifert, and which was the tallest building to have ever been demolished in the UK, in 2007

'The time to repair the roof is when the sun is shining' (John F. Kennedy, State of the Union Address January 11 1962)

The only certain thing in life is change - sometimes predictable, often not. The best you can hope for is that you are able to react and adapt.

The Hole-in-the-Wall Gang was a gang in the American Wild West, which took its name from the Hole-in-the-Wall Pass in Johnson County, Wyoming, where several outlaw gangs had their hideouts. No lawmen ever successfully entered it to capture outlaws during its more than fifty years of active existence (Wikipedia)

In around 1990 ATMs[29], or cash machines, were proliferating - there was something of an arms race going on between the major banks, and ATMs were seen as key measures of success, and one subject of "bragging rights"

I was then working in NatWest's Head Office, in Lothbury London EC2, and had to answer several press enquiries about how many we had. Work was intense at that stage, and there was no source of definitive numbers that I knew of. Once I had established that the caller had no idea what the true number was, and extracted from them whatever numbers they had obtained from other banks, I made up a number, and everyone was happy. They say that 32.7% of all statistics are made up on the spur of the moment. I tried to keep a jotted note on the corner of my desktop blotter of what I had said to whom and when, to help avoid any future embarrassment, but could rarely lay hands on it. I

[29] Automated Teller Machine

read now that in 1990 Midland Bank had 2,000 ATMs, and that Halifax Building Society had 1,200 in 1989. I suspect that both had a counterpart Bill doing the same thing as me, but we'll probably never know. What I do know is that ATMs had a very disruptive effect on branch banking.

They each cost over three times my annual salary at the time, but customers liked the ready, intimate private access to their money 24 hours a day without needing to actually go into the bank. And from the bank's point of view, they were cheap labour for a low-margin boring service, conducted on free real estate. That created some problems, with receipts left littering the pavement. Various designs of thin bins for unwanted receipts were attached to walls beside the machines, but they extended over the public highway where they were a pedestrian collision risk, and were being used for all litter, including by smokers as an ash tray so as well as requiring emptying they also represented a fire hazard. And they didn't even address the root problem. Receipts were then given by default, whether requested by customers or not. I suggested reversing this default, making disposal of receipts which customers had requested their own responsibility, and absolving us from blame for littering.

And ATMs were always breaking, back then, weren't they? Although they seem to have become far more reliable in recent years, I seem to remember Cash Machines were often out of commission. Perhaps it isn't all down to improved engineering?

Branches were graded according to the size of their business, measured in terms of funds on deposit, size of loan book, amount of insurance sold, staff numbers, service fees and so forth. And there was a career progression pecking order for the Manager and his management team, with associations between branch grades and the grades of staff assigned to each. The number and value of cashier transactions were included in the measures for each branch, but it was the sale of additional 'products' to customers which were of greater value. In the jargon of the day, that was referred to as enticing customers 'off the lino and onto the carpet'.

When ATMs appeared they were ascribed to a different, central cost centre, and branches were simply recompensed for refilling them with cash. That took the pressure off cashiers, but it also took customers further away from the carpet.

An ATM monitoring centre could see the volume and value of all transactions passing at each machine, nationally and, crucially, received alerts when cash was running low. Calls would be made to branches to alert them and to prompt re-filling. In practice, the machines often ran out of cash, and the branch manager's response was to say that staff were too overwhelmed with manning the (busier) cashier points to be able to refill the machine. By strange coincidence, those customers were also brought into the branch and closer to the carpet.

Making an adjustment to accounting practices, giving managers a sense of ownership and greater incentive to support the smooth operation of "their" ATMs, went a long way towards improving the "reliability" of the machines.

'No army can stop an idea whose time has come' (Victor Hugo)

Very rarely something big comes along and drives a seismic shift in everything. A World War, or the Industrial Revolution.

On 5th February 2022 Nadine Dorries, the UK Secretary of State for Digital, Media and Culture, informed the British public, live[30] on Sky News, that 'we've had the internet for 10 years'. But like most others, I knew that Tim Berners-Lee had "invented" the world-wide web in 1989 (and that anyone with a browser could learn, if they didn't know already, that that meant designing new protocols[31] to add functionality to the internet, which had been developing since the 1960's).

Despite what Ms Dorries would have us believe, the 'Web began as a publicly available service on the Internet on August 6, 1991, when Berners-Lee published the first-ever website[32].

[30] https://www.youtube.com/watch?v=XU5NgUClimc&t=919s

[31] The key technologies that underpin the WorldWideWeb ('the Web'), include Hypertext Markup Language (HTML), for creating Web pages; Hypertext Transfer Protocol (HTTP), a set of rules for transferring data across the Web; and Uniform Resource Locators (URLs), or Web addresses for finding a document or page.

[32] http://info.cern.ch/hypertext/WWW/TheProject.html

World Wide Web

The WorldWideWeb (W3) is a wide-area hypermedia information retrieval initiative aiming to give universal access to a large universe of documents.

Everything there is online about W3 is linked directly or indirectly to this document, including an executive summary of the project, Mailing lists , Policy , November's W3 news , Frequently Asked Questions .

What's out there?
: Pointers to the world's online information, subjects , W3 servers, etc.

Help
: on the browser you are using

Software Products
: A list of W3 project components and their current state. (e.g. Line Mode ,X11 Viola , NeXTStep , Servers , Tools , Mail robot , Library)

Technical
: Details of protocols, formats, program internals etc

Bibliography
: Paper documentation on W3 and references.

People
: A list of some people involved in the project.

History
: A summary of the history of the project.

How can I help ?
: If you would like to support the web..

Getting code
: Getting the code by anonymous FTP , etc.

The first-ever web page

In 1993, Mosaic, the first Web browser to become popular with the general public was launched, and the next few years saw the launch of Yahoo (1994), Netscape[33] (1994), Amazon (1995), eBay (1995) and Google (1998) (though few had heard of the last three mentioned newbie minnows in those days).

NatWest, where I was working in 1997, like most corporations at the time, was nowhere. That year they put up a page to let people know their address and contact telephone numbers. But mostly to be able to say that they were 'on the 'web'.

---

[33] Netscape Navigator quickly had more than three quarters of the web browser market.

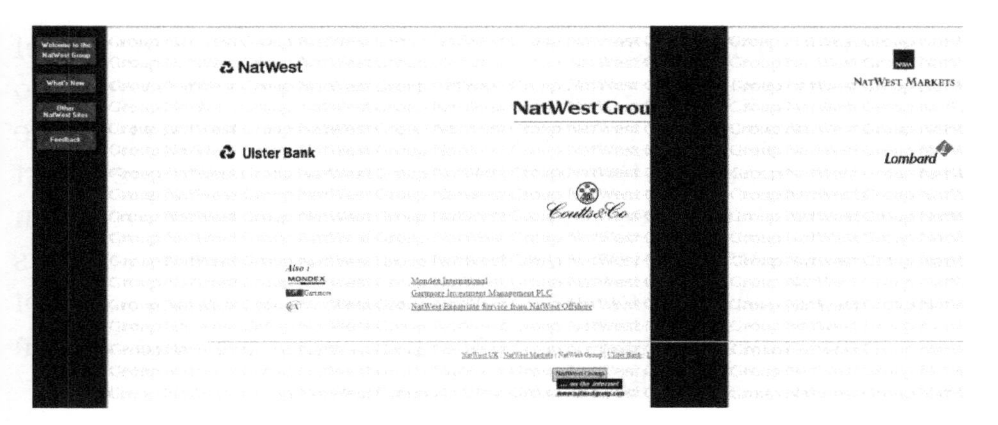

NatWest web site in 1997

At that time the first web search engines were proliferating – Infoseek, Yahoo, Webcrawler, Lycos, Excite, AltaVista and AskJeeves (but, as yet, no Google nor a Microsoft offering), whilst Netscape dominated the browser market, with their 'Navigator'.

The web was of particular interest to me because, at that time, I was responsible for IT support for a large number of NatWest business support functions, and was in the midst of the frankly anarchic heyday of 'distributed client-server computing'. Mine was a managerial, not a technical role – I have only a rudimentary knowledge of IT, not a Computer Science degree.

It was at a time when more and more work was being done on PCs which were attached to local area networks, along with networked printers and 'file servers', where files could be saved (or not) and backed-up (or not). Oh, and with email. The halcyon days of a few 'dumb terminals' connected to a central mainframe computer were becoming a distant memory, as more and more desktop PCs appeared, with a burgeoning variety of software. Think of it as a teacher trying to control a class full of mischievous children from the corridor outside the classroom.

It seemed to me that web technology offered a means of making information widely available and an opportunity to regain a degree of control, with a "single version of the truth"[34] – ensuring the availability to all of only the latest version, without needing to have control over or even detailed knowledge of the platform which the recipients were using. Although the new medium offered all sorts of potential, my interest at that time was for use on an intranet; solely for the benefit of bank staff.

Those were exciting, pioneering days, though nobody knew how far it would all go and what changes would result. There were just a handful of us at the bank who were interested and excited by the opportunities and potential and, with no mandate or sponsor, we had to work clandestinely and largely in our own time.

We looked around for the pace-setters, and quickly spotted that the BBC seemed to be leaders. They had everything going for them – the Web was the perfect medium for broadcasting - potential global reach without most of the physical infrastructure challenges. And they had plenty of experienced, creative people. Perhaps because we presented no competitive threat and were eager fellow travellers, they were generous in sharing experience.

Of particular interest to me was their use of the technology on an intranet, for knowledge- and experience-sharing, creative collaboration and production team-building. Specialist professionals, such as cameramen, costumiers, sound engineers and so forth, were each allowed a personal page to curate, where they could advertise their experience, skills, aspirations, and availability, so that producers could

[34] This seems ridiculously naïve from the historical perspective of the "post-truth" world of "alternative facts" which now plague us

readily view their internal talent pool and put together teams for new projects. It also meant that a large part of the substance creation fell to those with the greatest incentive, spreading the workload, and accelerating content creation. The development team were left to concentrate on creation of the platform, standards, content templates, and integration[35].

he BBC had launched their pioneering web site - bbc.co.uk - in 1997. It's often forgotten, now, but the BBC were in the top 10 most popular websites in the world for over a decade – from June 1997 until February 2008, peaking at number 5 between December 2000 and October 2001[36].

[35] In later life I reflected on the similarities of Apple's App platform, and the genius of a model which provided a huge incentive for third parties to develop, at their own cost, content that enhanced the allure of Apple's product.

[36] See animation by James Eagle, eeagli.com, 'Most popular websites since 1993' at https://www.youtube.com/watch?v=hNDILCdZmRo

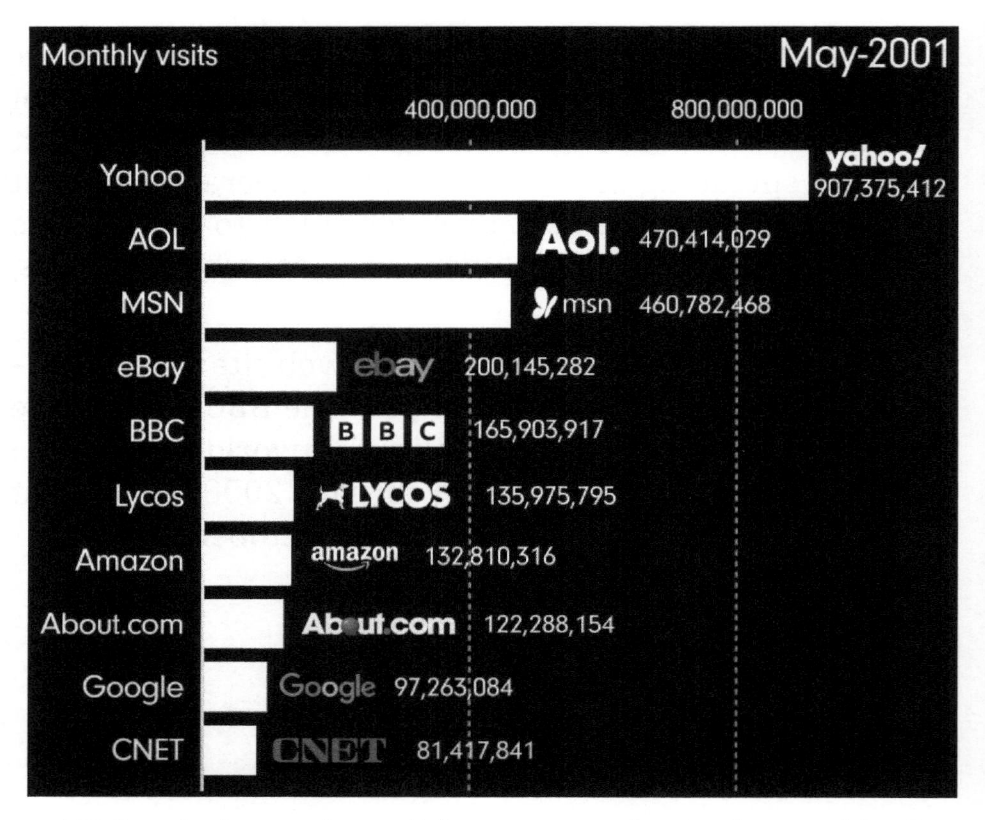

BBC web site peaks at 5th most popular in the world - From 'Most popular websites since 1993'

The BBC had launched Ceefax (a pun on 'see facts') - the world's first Teletext service - in 1974, back in the days when Britain spent time actually being world-leading in some things rather than simply talking about it. Using about 25 of the "spare" (625) screen lines to transmit up to 100 pages[37] of 24 lines by 40 characters on a loop repeated every 25 seconds. Almost 50 years on I still have fond memories of the availability and relative immediacy of information, such as weather, news, flight arrivals and racing results. It was the web of its day.

[37] There were only 24 pages, originally

P100 CEEFAX 1 100 Wed 22 Dec 16:24/41

BBC CEEFAX

History

KEY EVENTS FROM THIS DAY IN HISTORY 143

A-Z INDEX	199	NEWS HEADLINES	101
ON THIS DAY	143	CONTACT DETAILS	695
ENTERTAINMENT	500	NEWS FOR REGION	160
FINANCE	200	SPORT INDEX	300
SCI-TECH	154	TV LISTINGS	600
WEATHER	400	NEWSREEL	152
CRICKET	340	RUGBY UNION	370

Ceefax: The world at your fingertips

Headlines Sport N.Ire TV A-Z Index

The BBC CEEFAX 'home screen'

The BBC 'Networking Club' was, I think, running from late in 1994. Later, with John Birt's[38] support, they began developing a global news service on the web but, with Ceefax the mature information source of the BBC, their early work included a process to extract certain data from Ceefax via automated scrolling through screens displayed, (possibly capturing a 'screen-scrape' image of each, then running them through an OCR[39] procedure), and converting to standard text, before re-rendering in HTML[40]. Although it was a great

---

[38] BBC Director-General at the time

[39] Optical Character Recognition – converting an image of text into a digital understanding of the characters

[40] Hypertext Markup Language – one of the protocols which underpin the WorldWide Web, and what web browsers were designed to be able to read and display

example of reverse engineering, and something of which W. Heath-Robinson would have been proud, it worked, and allowed the team to repurpose the output from the Ceefax team in the short term, without the need for wholesale organisational change.

The BBC's web site in 1997

Over the next few years, whilst commercial exploitation of the potential functionality – including rapid news feeds, video, user-polling, and e-commerce - was frenetic, we were content to develop an intranet to enable the much simpler, faster and more efficient sharing of mundane information about support services, like stationery, company cars and legal services, to the bank's staff.

The rest, as they say, is history, and you can read all about the bigger picture on the, er, WorldWideWeb.

Business Architecture

'If you can't explain it to a six year-old you don't understand it yourself' (Albert Einstein)

In 2003 I stumbled into a role called 'Business Architect'. It felt like I had finally completed my career journey. Home. But I came to realise that it was the brief time which I had spent blagging a living in IT management that probably won me the job. It turned out that my new employer had chosen to pursue an 'Enterprise Architecture' approach, and felt the need for a Business Architect to help.

At that time the whole field was a kind of wilderness exploration - a lonely place, with very few "practitioners" from whom you could gather "best practices" or even share ideas. But on the plus side, working for a high profile but commercially non-competitive enterprise made thought leaders more amenable to collaboration, and the work was pioneering, challenging, worthwhile, and enjoyable.

Now is probably as good a time as any to point out that if you're looking for a text book or instruction manual about Enterprise or Business Architecture then you have probably bought the wrong book. There are plenty of those available – mostly dry and academic. You'll find that there are a wide variety of opinions. These are mine.

'The most basic question is not what is best but who shall decide what is best' (Thomas Sowell)

Firstly, let's establish the definition of an 'Enterprise'[41]. I suspect that the dictionary definition 'a company organized for commercial purposes; business firm' is the one most conventionally assumed in the term 'Enterprise Architecture'. I prefer to include the possible non- formally organised common endeavour, either for or not for profit, and 'a project undertaken or to be undertaken, especially one that is important or difficult or that requires boldness or energy'.

Although its roots lie somewhere in the 1960s, the term Enterprise Architecture (EA) was not coined until the end of the 1980s.

John Zachman is rightly regarded as its Father although, with characteristic modesty, he pays credit to Dewey Walker, Director of Architecture when they both worked for IBM, for his contribution. Walker shared his thinking on what he called 'Business Systems Planning' with Zachman, then an account executive, which helped him to deal with a major oil company merger. But it was Zachman who went on to develop the framework which bears his name.

Business travel can be a lonely experience, especially in foreign lands. I remember taking John for dinner in London during the time, in 2003 and 2004, that he was helping us. The waiter's mother tongue was Russian, and John seemed impressed that I was able to hold a rudimentary conversation with him in his native language. I was worried that I might have blown my cover. The next day I shared a podium with John presenting thoughts and plans

[41] First recorded in 1400–50; late Middle English, from Middle French, noun use of feminine of entrepris (past participle of entreprendre "to undertake"), from Latin inter- inter- + prehēnsus, prēnsus

for implementing an Architecture Framework based on his model.

The Zachman Enterprise Architecture Framework (ZEAF) emerged in 1987 and is, he says, 'just the definition and structure of the descriptive representation for enterprises'. It is an ontology for EA, based around the six key questions What, How, When, Who, Where, and Why - not how to do EA.

Zachman's Framework describes a route for transformation of the strategy into the operating Enterprise, comprised of very precise instructions (explicit or implicit) for the behaviour of people and/or machines.

I agree with John Zachman, when he says that 'Enterprise Architecture is not about building IT models - it's about solving general management problems' but, although most practitioners today claim that EA is more than just high level systems engineering, the fact is that most enterprise architecture practitioners report to a CIO[42] or other IT department manager. The work of most EA teams is to help business and IT managers develop strategies to support and enable business development and change in relation to the information systems that the business depends on.

Business Architecture grew out of EA to, I believe, bridge the gap which had been allowed to develop, or at least to forge the connection between, business and IT. Although many corporate bodies have for decades had on their board a Chief Information Officer CIO), in recognition of the important role of information in businesses today, in my experience there remains a disconnect between IT and the

---

[42] Chief Information Officer

business – particularly at levels below the CIO but within her purview.

My own view is that there are generic business-specific skills and techniques, which can be viewed through an architectural lens, that can assist the connection to the world of IT but which can also offer value independently. Not all solutions to business problems involve IT. Remember the "faulty" ATMs?

And there are, I believe, some fundamental differences between EA and Business Architecture. For example, rules are very important in systems engineering - IT systems can be designed to follow rules. But people and, in some respects, corporate entities do not follow rules. At best they pay attention to consequences, which may or may not be associated to rules, and to incentives. Perhaps it is because Enterprise Architecture addresses the world(s) of hardware and software whereas, in my view, 'wetware[43]' carries at least equal status.

There's plenty of history and, although you'd be forgiven for believing otherwise, no authoritative definition of the term and practise "Business Architecture" over others.

The American organisational theorist William R. Synnott presented one of the first models of business architecture in 1987, in the context of data management. He wanted to develop an overall Information Resource Management (IRM) architecture, and proposed business architecture as its foundation.

His IRM architecture model distinguished seven types of architecture components:

---

[43] Essentially, people

‘Centralized’: Business architecture, Data architecture, and Communication architecture, and

‘Decentralized’: Human resources architecture, Computer architecture, User-computing architecture, and Systems architecture.

But, although there were several proprietary software and methodology vendors there was no commonly agreed standard definition of what the term meant, or what the discipline included. There still isn’t, but those of us trying to do a job had to find something of substance to start with.

A core principle of mine has always been to use something which already exists if possible, or modify and reuse something, leaving the creation of new things to a last resort. I have also always sought to use open standards, from academia and non-commercial parties.

Back then ‘OMG’ had an entirely different meaning to that of today. The Object Management Group (OMG)[44], an international, open membership, not-for-profit technology standards consortium, was founded in 1989 and fitted the bill well. Several “standards” which they went on to lead proved useful to me in my work (more later) and, in 2007, they founded a working group called the Business Architecture Working Group (BAWG).

They defined the term “business architecture” as “a blueprint of the enterprise that provides a common understanding of the organisation and is used to align strategic objectives and tactical demands. It describes the

---

[44] The Object Management Group® (OMG®) is an international, open membership, not-for-profit technology standards consortium, founded in 1989. OMG Task Forces develop enterprise integration standards for a wide range of technologies and their modelling standards include the Unified Modeling Language® (UML®) and Model Driven Architecture® (MDA®)

structure of the enterprise in terms of its governance structure, business processes, and business information and the profession of business architecture as being primarily focused on the motivational, operational, and analysis frameworks that link these aspects of the enterprise together".

That's a pretty good place to start. The OMG went on to explain that "the key characteristic of the business architecture is that it represents real world aspects of a business, along with how they interact. Products of this business architecture work are used to develop plans, make business decisions and guide their implementations."

"In practice, business architecture represents a business in the absence of any IT architecture while enterprise architecture provides an overarching framework for business and IT architecture."

That closely matches my own view, and the approach which I have followed throughout my career. Frameworks and methodologies can provide useful guides and, occasionally, supports, but following any one of them slavishly inhibits the flexibility and creativity needed to focus on particular enterprise characteristics and meaningful challenges. 'Doing' EA or BA for its own sake, or with too much focus on 'methodological purity' at the expense of business purpose is, in my opinion and experience, likely to make the effort irrelevant.

For more than three years after my last post I was bombarded with invitations to apply for jobs with the job title 'Business Architect'. It seems, from the job market, that there has been some regression, to make the role once more tightly bound to the world of IT. Most 'Business Architect' job descriptions called for IT qualifications and skills. Occasionally they specified additionally knowledge or expertise in a specific business or business function.

Here, to illustrate the point, are a couple of typical recent Business Architect job descriptions[45]

Business Architect Job Description

Description

In this role, you leverage both your strong understanding of technology and of business management to define end-to-end solutions for the company. You demonstrate leadership in terms of looking beyond the pure technology aspects, considering the value that can be created by technology for the company, and changing how technology is viewed in the organization. You are adept at interacting, communicating and partnering with other departments within the company.

Key Responsibilities

- Define and maintain the Business Architecture
- Maintain the Business Capability Model and Process Catalogue
- Define and maintain the Business Principles Catalogue
- Chairing the Business Design Authority
- Support the development of the Target Operating Model
- Support Business Impact Assessments for change initiatives
- Act as a cross-functional leader to craft executive-ready business strategies
- Develop papers articulating the value of change proposals.
- Manage relationships with other departments and build effective enterprise best practices.
- Lead design thinking/digital transformation workshops with senior leaders.

Basic Qualifications

- 5 years of experience in Business Architecture role
- Degree in Computing and Masters in Business Administration (MBA) preferred
- Experience in value engineering, strategy consulting and/or enterprise architecture
- Experience in architecting, designing or delivering digital transformation programs
- Subject matter expert in of one or multiple parts of the IT stack or development coding platforms
- Subject matter expert in one or multiple lines of business
- Ability to build financial models and quantitative/qualitative analysis
- Excellent verbal, written, and formal presentation, communication, and facilitation skills.
- Knowledge of value chain analysis, and benefits realization of technology investments

[45] Yes, they are taken from authentic ones, which I have modified merely to genericise and anonymise

Business Architect Job Description

Description

In this role you will be responsible for providing as-is and target operating models for a range of back-office business operations, from sales and marketing to finance and HR

This role supports the Solution Design Authority and Systems Transformation Office in ensuring that business change brought about by the successful delivery of systems transformation is **understood and accepted by business process owners**

Key responsibilities

- Manage and create as-is and to-be operating models for back-office business functions such as sales and marketing, finance and HR, etc
- Be responsible for ensuring business analysts and implementation partners are documenting requirements aligned with the target state operating model
- Provide support and guidance to analysts in the capture of business and systems requirements
- Engage in and facilitate business architecture and design workshops
- Ensure the impact of changes on the operating models are understood and reflected against business cases, to ensure solution decisions can be made in the context of the business operation
- Manage relationships between business function owners to ensure understanding of business models and any associated change
- Prepare and provide regular reporting (progress and status) to key stakeholder groups in accordance with required format and timeframes

Required qualifications

- Experienced working in a senior management role in a large corporate setting.
- Experience of working on major systems transformation projects involving business architecture solutions design and delivery.

Notice the IT slant in both. In the first the role is responsible for 'changing how technology is viewed in the organization' and in the latter for ensuring that business process owners 'understand and accept' systems changes.

The position is usually recruited from within or at the behest of the IT department – often because they have little or no experience of, real interest in or care about the real business world. A cynic might say they are looking for someone or something to give their work justification. In my experience, most businesses that "use" Business Architecture do so in the same way that a drunk uses lamp-posts - for support rather than illumination[46]. In other words, they use it to demonstrate a connection between an

[46] to paraphrase Andrew Lang

IT or other driven project with agreed corporate goals or priorities.

Contrast those example Job Descriptions with this view from CIO.com[47]

> If I were hiring a business architect, here is what I would look for:
>
> **A sound understanding of business principles and concepts.** Most IT types think understanding the business is all about understanding the business processes but this is not what business leaders are interested in. Business architects should understand how the market context affects the business, how value is created, what differentiates their company from its competitors, and how products are created, marketed, and sold. They should have a good understanding of how business strategy is developed – even if it is never articulated.
>
> **An ability to think about business processes outside of the technology context.** Even business people have a hard time with this. I have had more than one business architect share his frustration with business project people who continually talk about business process in terms of how their applications work. Though a business architect needs to understand how to leverage IT for business value, he needs to be able to draw a wide, heavy line between business processes and the technologies that enable them.
>
> **A really strong consulting mindset.** Building a good business architecture is more about listening and hearing between the lines than selling a concept or framework. At the end of the day, the successful business architecture will be one that business leaders resonate with. Business architects should see themselves as business consultants looking for problems to solve.
>
> **Have a strategic point of view.** Business architects need the ability to challenge people's thinking; to get them out of the current issues and current systems and into thinking about the possibilities of the future. As one of my clients so eloquently put it: "It's not so much about thinking outside of the box as it is thinking outside of YOUR box".
>
> **Good at design thinking.** I want a business architect that can bring order out of the typical strategy chaos of most companies. That means that they can listen to lots of ideas and create a view that resonates across a wide part of the organization. It means they have the ability to see what others are missing and can create a clear line of sight between business intention and business action.
>
> **A catalyst for change.** At the end of the day, business architecture isn't worth the napkin it is scribbled on if the organization doesn't change. A business architect should see himself as a change agent fist and an architect second. He should use business architecture as a tool to agitate for action

This is much closer to my own view.

---

[47] 'Six Attributes Every Business Architect Should Display' https://www.cio.com/article/2372660/six-attributes-every-business-architect-should-display.html

I believe it can, if properly designed, be at the very heart of the Enterprise, and the key tool to driving those priorities. It can, for examples help

- maintaining enterprise-wide focus on corporate objectives and alignment with principles;
- Identification and prioritisation of change initiative candidates - providing a yardstick for considering disparate options against a common set of measures;
- Making new staff informed and aligned to corporate values and goals more quickly and consistently
- Undertaking predictive analysis - to help assess the effect of proactive change in one element on others;
- Enable more rapid assessment of the effect of external changes;
- Sharing of good practices and retention of corporate knowledge;
- Make it easier to identify gaps and overlaps of elements, so as to focus resources more efficiently and avoid duplication;
- Providing the basis for designing changes, to include IT systems, training, guidance, responsibilities etc;
- Reviewing outcomes against planned expectations;
- Contingency and business continuity planning.

Most importantly, and fundamentally, Business Architecture helps to clarify the intent of the Enterprise and to chart a path, through strategy, to realising that intent. It is also about solving problems or, better still, avoiding them in the first place, and about spotting needs or opportunities offered by change rather than just helping to respond to it.

Business Architecture in my view helps clarify the high level landscape, scope, and strategy, and provides a framework and a series of tools for others - such as Business Analysts and Programme Managers - to work with and within.

Just as the discipline of building Architecture begins with the purpose of a building, describing its various parts, functions and the relationships between them, in clear, structured terms, for all its users, so a documented 'Business Architecture' aims to do the same for an Enterprise. It may be thought of as a blueprint for those needing to undertake maintenance or modifications, or simply needing to ensure that the most efficient use is made of the skills and other resources at its disposal.

Another benefit of separating elements into distinct parts is that each may be independently considered and managed whilst systematically maintaining the relationships between them.

By distilling the complexity of the Enterprise's operational arrangements into their constituent, elemental parts and the relationships between them, when a change to any part occurs or is being considered, the effect on all of those other parts with which it is related may be identified[48]. So, for example, when new legislation which affects the Enterprise comes into effect, all processes that it governs, the roles that are responsible for them, resources used or consumed, Strategies and Goals that they aim to achieve may be considered.

It can represent a central point of reference and a corporate asset for the Enterprise and is intended to be an evergreen model, developing over time in scope and completeness in

[48] So long as you have a Metamodel to maintain the connections – see later

response to operational needs and with the benefit of practical experience.

You can use as much or as little as you feel you need. I believe that there are some parts that every Enterprise needs and should create and keep close to their hearts at all times, and I don't think it's necessary or cost-effective to develop all areas in detail. Use as much or as little as you feel you need. Most of it won't ever be used, but there will be some parts where important issues hinge on points of detail, and the architecture can provide a framework within which the needed detailed analysis can fit. Think of it as a Town Plan, where you need to define key functions, functional zones and arterial transport links, but you don't actually need to build most until it's needed.

But there aren't any rules, and I'm not the Business Architecture police. Actually, there aren't any Business Architecture police, and you shouldn't believe anyone who tells you different.

As with any model, a Business Architecture model is just that - a simplified representation of reality and a device to aid comprehension and communication. And keep in mind the aphorism 'all models are wrong, but some models are useful'[49].

You might decide that you already have all or most of this sorted and clear in your or someone else's mind[50]. If that's the case, you ought to be concerned, or at least take out generous 'key man' insurance. Maybe it's all written down in a variety of disparate places? It ought to be quick and

[49] George Box, 1976

[50] Individuals clutching knowledge which only they hold can represent the seat of their power and sense of importance, and helps to explain why they are frequently reluctant to give it up.

easy to fill in the gaps and to tie it all together, for the benefit of everyone in the Enterprise, and the Enterprise itself.

At the very least Enterprise and/or Business Architecture can offer a method for referencing or cross-referencing those disparate information sources. Without a high level view of the whole range of information, however, it won't be possible to make the connections and associations that a Business Architecture perspective can provide.

Business Architecture is much more than just documents. For me, it's a way of thinking and, like Euclidean Geometry[51] it can be of great practical value, so I am going to talk about it as a discipline or practise in general.

Finally, whilst entropy might be just what a new, or start-up enterprise needs, I recommend that you start from the presumption that things are the way they are because it works. Try to resist the temptation to plough in and meddle just for the sake of it - at least until you see evidence of specific need. Any mature enterprise will have had time to evolve, and settle down into a steady state. However inefficient or even "clunky" you might think it, the fact that it is still here at all, in the face of Darwinian evolution suggests that it is a winner, or at least a survivor.

---

[51] Euclid's book, "The Elements", is a collection of axioms, theorems and proofs about squares, circles acute angles, isosceles triangles, and other such things. The five postulates (axioms) are: 1. Any two points can be joined by a straight line. 2. Any straight line segment can be extended indefinitely in a straight line. 3. Given any straight line segment, a circle can be drawn having the line segment as radius and an endpoint as centre. 4. All right angles are congruent. 5. (Also called the parallel postulate.) If two lines are drawn that intersect a third in such a way that the sum of inner angles on one side is less than the sum of two right triangles, then the two lines will intersect each other on that side if the lines are extended far enough.

Models are vital tools to help clarify complex concepts, and I often find diagrams helpful. I'll use plenty - mostly my own – sometimes accompanied by an explanation of how they were arrived at and what they may be useful for. But not everything can be best described by a diagram. I never found any satisfactory way to model the culture of an organisation, for example, though it's often vital to be aware of and bear in mind, especially when two organisations, or even departmental "bubbles", with different internal cultures need to collaborate closely. In most large organisations, think of the internal and professional cultures of Human Resources, Marketing, and Accounts departments, for example.

Fundamental organisational culture "tells" could be picked up very quickly in the receptions of virtually every police HQ that I visited, for example. In nearly all, a Force Roll of Honour (to the fallen) book (always open at the current date), and a packed sports trophy cabinet featured prominently. I never had the opportunity to witness the equivalent indicators in Social Services' offices, but I suspect that they were very different, and I sensed a tension between the two professions when engaged on necessarily joint operations.

As I said, earlier, use as much or as little of what follows as you need. Don't do any for the sake of it, but there are some basic elements which I believe are fundamental, and essential. The good news is that they are so basic they ought to be easy to complete. If they aren't easy, then take it as an indicator that you need to think about them, and going through the creation process might throw up a few unexpected insights, and will likely prove cathartic.

Firstly, it's helpful to begin by getting a clear sense of place –the context in which your enterprise operates, its principal role, interfaces and boundaries.

I created this diagram to help clarify the role that the police played in criminal justice [52]:

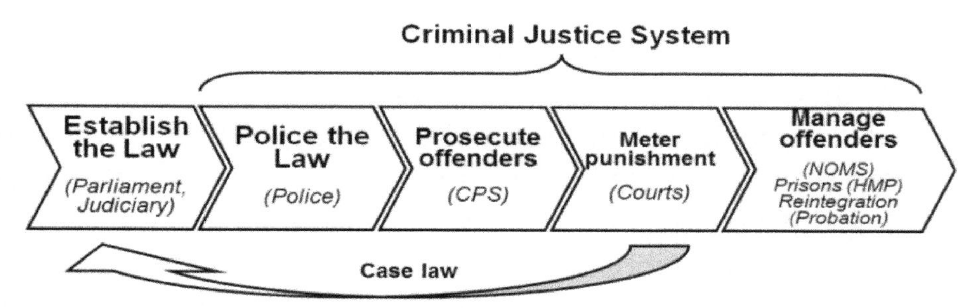

You then need to identify key attributes which might help you to understand what you later find. For the police service, for example, I found that

- They operate as part of the criminal justice, public protection and other public service 'supply chains', where each participant needs to at least understand each other's goals, methods and so forth in that particular 'chain', have slick hand-offs and integration where necessary[53].
- Traceability is vital – for detection, but especially for successful prosecution[54], so evidential integrity and

[52] The police also operate in other contexts, such as public safety and protection

[53] I learned, for example, that firemen use their whistles to warn colleagues of risk which they should hurry from, whilst a policeman's whistle called for assistance – not the kind of confusion you need in a smoky environment

[54] Known as the 'Golden thread' – reflecting the duty on the Prosecution to prove a prisoner's guilt beyond a reasonable doubt

the rigorous following of rules dictated by legislation[55] is vital

- The principles that guide their behaviour are long-established and clear (see later)
- They are fiercely loyal, and particularly supportive of colleagues
- Police forces operate most like regional franchisees. That wasn't a popular conclusion (I think it was taken as comparing them to MacDonald drive-throughs), but simply recognised that they operate under the same legislative rules, on a broadly common model and with a common purpose, wearing similar uniforms and with common vehicle livery, but with local prioritisation. But the public, and other agencies with whom they work (usually organised by different geographical boundaries) ought to be able to expect consistency from them.

[55] Particularly the Police and Criminal Evidence Act (PACE) 1984, the legislative framework for the powers of police officers in England and Wales to combat crime, which provides codes of practice for the exercise of those powers

Thinking about particular challenges, or anomalies, can also be informative. Imagine, for example, there's a debate raging about there being too much crime and/or too little punishment. Here's a diagram which I created in 2005/2006, based on the data at the time[56], when just such a debate was underway.

The bottom line 'only 2.5% of offenders are punished', seems to give the lie to the claim that 'crime doesn't pay', and makes a compelling headline. A cacophony of angry sound-bites and often politically motivated press headlines usually follows particularly heinous incidents, and a knee-jerk call for tougher sentencing.

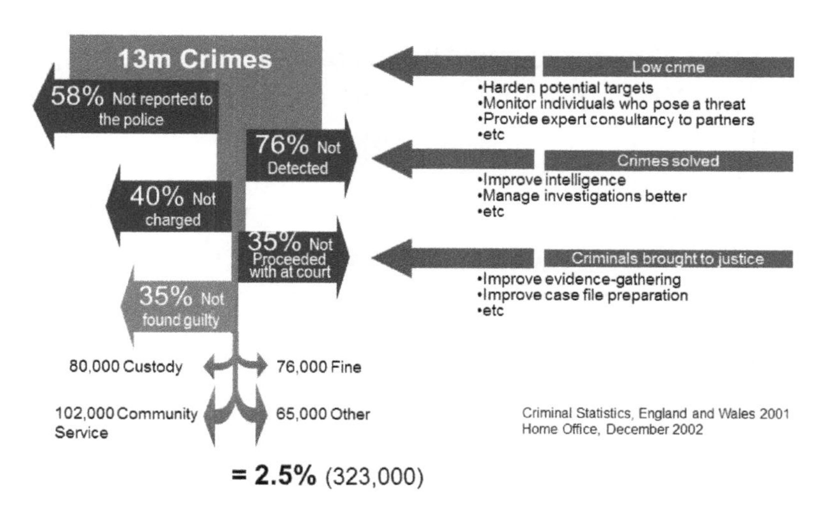

[56] There have been some changes, like in how data is gathered for the British Crime Survey, and most recent statistics were skewed at the time of writing, due to the coronavirus (COVID-19) pandemic and associated lockdowns. But the general scale and proportions remain similar and relevant for the purposes discussed here. Latest data can be found at www.ons.gov.uk/peoplepopulationandcommunity/crimeandjustice/bulletins/crimeinenglandandwales/yearendingseptember2020 , and I note that there remain concerns about the quality of recording by the police, and that crime is not recorded consistently across forces.

There is plenty here to find disturbing. There are about 13m crimes per year in a country with a population of about 65m - suggesting that, on average, you'll be a victim every 5 years. For balance, there were about 6m people on the PNC[57], suggesting that one in 11 people were the perpetrators.

Or that such a low proportion of offenders are punished. It may be, of course, that not all of the '13m Crimes'[58] were crimes that you are bothered about, but that still leaves a lot of 'perpetrators' unpunished.

I think we can agree that the true goal ought to be to reduce the incidence of crime. As well as improving the general sense of wellbeing in the community, it will result in great cost savings later on, as the demand on the (expensive) criminal justice, and prisons and rehabilitation services is reduced.

So, starting at the top of the diagram – what can be done to reduce the number of crimes committed? Let's consider the challenge in a structured manner.

Practical measures like 'hardening targets' (improving home security, bicycle security-marking, improved street lighting, public advice campaigns, 'Neighbourhood Watch' schemes and so forth can help, as can improved collaboration between and providing consultancy to interested agencies, such as schools, hospitals, shops, Local Authorities.

---

[57] Police National Computer – records of those CONVICTED of crimes

[58] as reported through the British Crime Survey

Improvements to car security over recent decades have resulted in a significant reduction in car crime[59], and more could be done, such as making dash-cams, trackers and 'black boxes' compulsory on all new cars. The proliferation of CCTV has already, undoubtedly, had an effect not only on crime solving, but discouragement and, again, quality and functional developments[60] could be specified and encouraged through commercial insurance rate incentives.

Higher conviction rates would also likely have a deterrent effect, and therefore result in a reduction in the number of crimes committed. A virtuous circle. As I said previously, people do not follow rules, but pay attention to consequences, and it's reasonable to assume that at least a proportion of crime is based on a risk/benefit assessment in that consideration. Is the potential benefit worth the risk? The sentence tariff will be one factor on the risk side of the equation relating to the commissioning of many crimes, but so too will the likelihood of being caught then successfully prosecuted and sentenced. And if the chance of that is vanishingly small, then increasing tariffs will have little effect. Conversely, if the chances of being successfully prosecuted are dramatically increased, and the risk is increased correspondingly, it would be reasonable to expect an overall reduction in crime. Which must, surely, be the principal aim?

Of course not all crime is carefully considered in advance by perpetrators, but it's worth analysing in detail from this perspective the cause at each point of "attrition". Further

---

[59] As an example of how insecure cars once were, I once had a car which could be started using a wooden ice lolly stick in the ignition.

[60] Evidential Image quality, guaranteed time-stamping, speed of search, retention regime, integration for examples, to improve deterrence, detection efficiency and increase the rate of successful prosecutions

analysis from different perspectives (e.g. economic desperation-driven or drug-induced crime) would probably follow.

So next up - why are less than half of crimes reported to the police? Police-recorded crime excludes less serious offences dealt with by magistrates' courts (for example, motoring offences), but it does have a wider offence coverage and population coverage than the (now telephone-operated) Crime Survey for England and Wales (TCSEW). On the other hand, a proportion of property crimes are only reported in order to secure a 'Crime Number' in order to be able to lodge an insurance claim, and an even greater proportion aren't reported at all, principally due to low expectation of a successful prosecution, or a perceived risk from perpetrators.

That reduces the numbers by an astonishing 58%. The police can't solve specific crimes they're not aware of.

Next, improve detection rates. The police are already doing their best, and there have been innovations such as mobile ANPR[61] and personal identification technology, and body-worn video which been especially helpful in discouraging crime, dealing with incidents, supporting the gathering of evidence, and in support of prosecutions. Dictating changes to new car specifications and commercial CCTV systems, and improved collaboration with other agencies, as mentioned above, could also prove helpful in detection.

Next, the decision as to whether or not to prosecute. In the UK this decision is taken by the Crown Prosecution Service

---

[61] Automated Number Plate Recognition system, which is linked to other databases, such as motor insurance and wanted alerts, triggering flags on identification

(CPS), whose remit is to prosecute cases which have a 'realistic prospect of conviction'. A casual observer might assume this to follow the dictionary definition of 'prospect' as 'the possibility that something will happen'. And, since it is not the role of the CPS to determine guilt or otherwise, but merely to bring cases for the decision of the courts, one might reasonably expect the threshold to be something over about 35-50% which, to my mind, represents a 'reasonable prospect' of a conviction.

Much has been written about this, and it turns out that 'a realistic prospect of conviction' is taken to mean 'that an objective, impartial and reasonable jury or bench of magistrates or judge hearing a case alone, properly directed and acting in accordance with the law, is more likely than not to convict the defendant of the charge alleged'. This is a far lower test than the standard of proof required at trial in a criminal court because for a jury or magistrate(s) to convict a defendant in court at trial the prosecution must prove its case beyond reasonable doubt, and suggests an element of "pre-court judgement" (i.e. the Courts can only judge cases that are presented to them) by the CPS. So, 55-60%? It turns out that the proportion of successful prosecutions brought by the CPS is actually about 80%[62], suggesting they only take forward cases that they are fairly confident they'll win.

[62] From www.cps.gov.uk 77.9% overall convictions in Q1 20/21, down from 84.6% in Q4 19/20

Next I want to turn to developing a general structure which can be applied to any Enterprise. Think of it as a template for the blueprint which I mentioned earlier.

Documenting the Business Architecture of an enterprise

Firstly, language. Most of us understand basic, common English, and have access to a dictionary and online resources to help. But in Business Architecture precision and clarity are vital. Ambiguity with language can make clarity of thought difficult. Consider the simple phrase 'Give a sharp nudge to the nut holding the steering wheel'. Will that be the 5/16" UNF nut, or the driver? That English often has a number of synonyms for similar or nearly similar things only makes things more complicated. And words are merely labels or tokens for concepts or images. It is the concepts that are most powerful.

Many industries have their own terminology – vocabulary - and so do most enterprises; their own "internal language". Sometimes that language remains internal, often it needs to be shared with suppliers or collaborators. Sometimes different terms are used for the same thing with different audiences, and for various reasons. For example, 'aqua' and 'sodium chloride' are often shown as major ingredients of shampoos - presumably to obscure from customers of salon brand products their less exotic-sounding names, water and salt.

Of particular relevance for Business Architecture are the production processes which transform inputs into products and the rules and criteria that govern, determine or define them. Think of the brewing process, for example:

Starting with the basic ingredients of water, malted barley, brewer's yeast and hops, the processes of malting, mashing, lautering[63], boiling, fermenting, conditioning, filtering are performed in a variety of imaginatively-named specialist equipment, including a Mash Tun, a Brewing Kettle (or Copper), and a Hop Back, before being stored in Kegs/Casks or bottles. During that magical process the ingredients are transformed via Wort[64] into beer. What measures and conditions govern the start and end points of each?

And that's just the production process. The Business Architecture, and associated Glossary, will need to include all the terms relating to the Enterprise's marketing, procurement, testing, costing, sales, staffing, financial management, distribution etc.

Consider also that individual people or things can simultaneously fulfil more than one named "role". A bank Customer can also be an ex-'Young Saver', and a 'Potential Customer' of bank insurance or credit card services.

Complexity can breed confusion. There needs to be consistency and a shared understanding of terms, so that everybody within the enterprise knows what is being talked about in each context, and for that reason it is worth investing the effort to create and use systemically an Enterprise Glossary.

---

[63] the separation of the wort (the liquid containing the sugar extracted during mashing) from the grains

[64] the liquid extracted from the mashing process, containing the sugars that will be fermented by the brewing yeast to produce alcohol, and the complex proteins that contribute to beer head retention and flavour.

Finally, Business Architecture has some terminology of its own. You will have heard many of the terms and, by now, you shouldn't be surprised to learn that there's no agreed understanding of what they mean in this context, but there is reasonable consensus.

It can be helpful to produce a matrix of things which are of interest to the enterprise and the relevant states that they can be in. Transitions between them will usually be either proactive processes, or adverse events. That all sounds very complicated, so hopefully this simplified example based on winning banking customers and increasing sales of services to them will make it clearer.

To \ From	Individual not a Bank customer	Target customer of the Bank	Applicant customer of the Bank	Basic customer of the Bank	Customer of Bank savings products	Customer of Bank insurance products	Online Bank Customer
Individual not a Bank customer				Analyse cause(s) 'win-back' marketing	Analyse cause(s) 'win-back' marketing	Analyse cause(s) 'win-back' marketing	
Target customer of the Bank	Marketing						
Applicant customer of the Bank	Customer initiative Assess against criteria	Assess against criteria					
Basic customer of the Bank			Sales Account opening procedures		Analyse cause(s) 'win-back' marketing	Analyse cause(s) 'win-back' marketing	Analyse cause(s) Promote o/l banking
Customer of Bank savings products				Marketing Sales Docum'tion procedures			
Customer of Bank insurance products				Marketing Sales Docum'tion procedures			
Online Bank Customer				Marketing Sales Docum'tion procedures			

You can use the technique to map as many or as few areas as you need, in whatever level of detail is helpful. Each intersection will have an associated set of governing rules, and detailed processes.

Dark-shaded intersections are either impossible, rare or of no interest. Light-shaded intersections are events – typically unwelcome ones, calling for reactive action. Unshaded-intersections are proactive processes.

So, let's look at the Business Architecture in a logical order, when you'll see those terms (and concepts) in context.

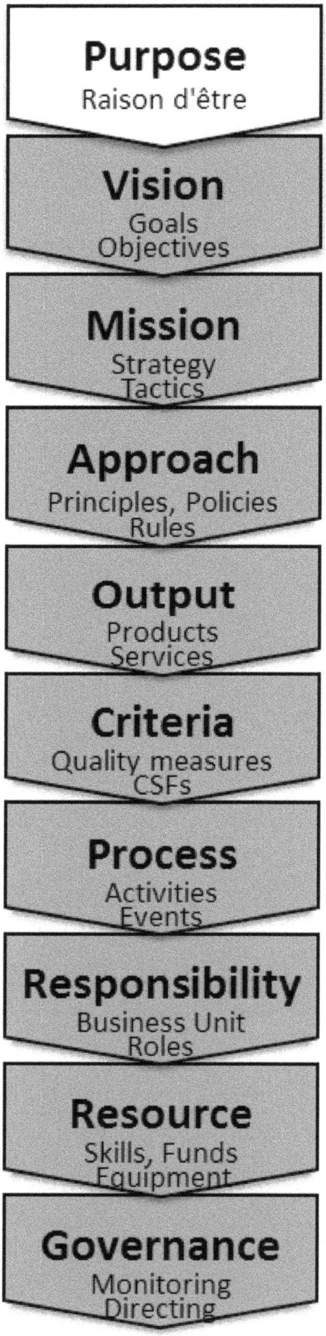

Purpose
Raison d'être
Vision
Goals
Objectives
Mission
Strategy
Tactics
Approach
Principles, Policies
Rules
Output
Products
Services
Criteria
Quality measures
CSFs
Process
Activities
Events
Responsibility
Business Unit
Roles
Resource
Skills, Funds
Equipment
Governance
Monitoring
Directing

Purpose

The first step must be to understand the basic premise of the enterprise. Why does it exist?

This model works for any Enterprise engaged in the production and/or sale of products and/or services, and has served me well throughout my career. It is vital that the nature of the enterprise and its principal motivation is understood, whether or not its products and/or services share shelf space with competitors.

The aim of a profit-making company is to make a profit. Sole traders and partnerships exist firstly for their own benefit, however nice they may be as individuals.

The aim of a limited company is to increase the value of the company for its shareholders. It is the legal duty of the directors to achieve that. How that is done will vary, as will what is done with the profit (such as payment to shareholders of dividends, or reinvestment in the company).

Charities and other 'not-for-profit' enterprises are also relatively straightforward, with a couple of complications. They need to be clear what their key purpose is, such as animal welfare, or improving the environment. But, even though it isn't their principal aim, some money needs to be made – to reinvest or to provide a buffer against leaner times and ensure continuity.

There are various other forms of enterprise which do not fit readily into any of these categories. Examples include

Building Societies, the largest of which is currently the Nationwide[65], who are Mutuals, owned by its members - anyone who banks, saves or has a mortgage with them - and run for their benefit. The Co-Op[66] is similarly owned by millions of members, but their principal business is food retail (they're the UK's fifth biggest). Bupa[67], operating in the healthcare arena, is a 'Provident Association' and, although a private company limited by guarantee, has no shareholders and reinvests its profits. The John Lewis Partnership[68] - owned by its staff, is just that, as is its supermarket, Waitrose[69]. Primula [70] – that delicious squeezable cheese – donates every penny of profit to charity.

Anything else can probably be best considered a hobby.

---

[65] https://www.nationwide.co.uk/about-us/what-membership-means
[66] https://www.co-operative.coop/about-us/our-co-op
[67] https://www.bupa.com/our-bupa/our-purpose
[68] https://www.johnlewis.com/customer-services/about-us
[69] https://www.waitrose.com/ecom/content/about-us
[70] https://primula.co.uk/love-to-share/

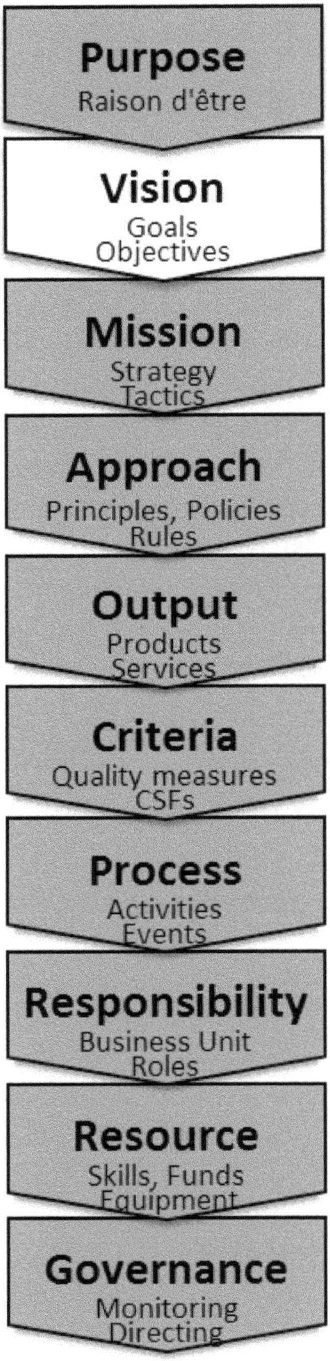

Purpose
Raison d'être

Vision
Goals
Objectives

Mission
Strategy
Tactics

Approach
Principles, Policies
Rules

Output
Products
Services

Criteria
Quality measures
CSFs

Process
Activities
Events

Responsibility
Business Unit
Roles

Resource
Skills, Funds
Equipment

Governance
Monitoring
Directing

Vision

Where do you want to get to? How you want things to be. How will others view the Enterprise – competitors, customers, posterity?

It should be expressed in static terms.

It is likely to already exist - often contained within political manifestos or White papers, investor briefings, and so forth, but is frequently quite woolly (suggesting that it isn't well understood. Now is a good time for some crisp clarity).

Vision can be broken down into contributory Goals and specific Objectives, each of which should contribute to the overall Vision, and be based on what needs to be achieved. You need to understand how success (or failure) will be measured, and an Enterprise Balanced Scorecard (more details later) will provide focus and a consistent yardstick of what is good and bad

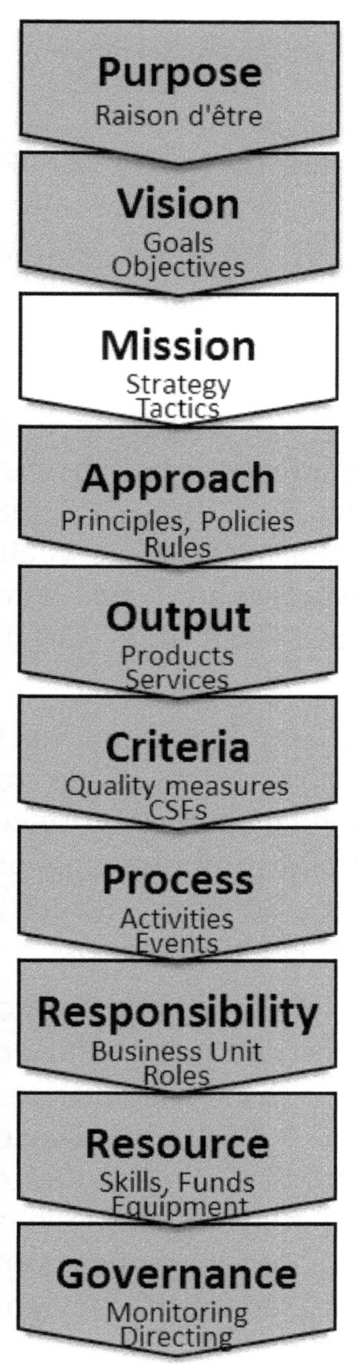

Mission

The Mission recognises the gap between the target (Vision) and current states – that you aim to bridge.

So it is about intended action – what you need to do to achieve the Vision - and should be expressed in verb terms.

Mission should be broken down into contributory **Strategy**, describing how the organisation will go about its tasks, and **Tactics** that it will employ.

Here's where you can recognise that it's unlikely that a single transition from current to target is possible, so define stages - how much specific things will be changed, in what order and over what timescales.

Approach

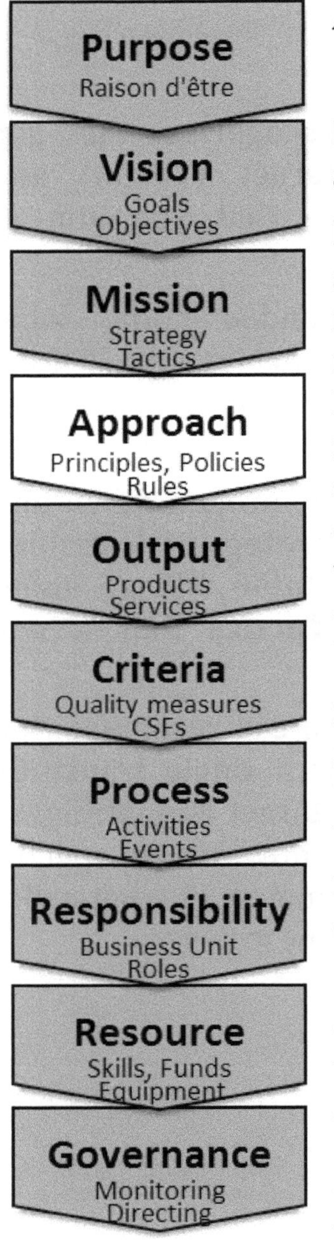

Next, define the way in which things are to be done.

Principles are fundamental statements of fact that serve as a foundation for decisions and policy formation, and act as a beacon for the Enterprise. They are enduring and inherent in the set of values held by the enterprise (you'll often see organisations telling you what their values are). They are what it needs to occasionally remind itself of, or to fall back on.

Specific **Policies** and **Rules** support and should be directly relatable to Principles. They give them substance by directly controlling, influencing or regulating actions. Rules are directly enforceable, whilst Policies are not.

But remember - people and commercial entities pay less attention to **Rules** than to the consequences attached to them. As (Lord) David Puttnam observed[71] 'Corporate responsibility and it's imposition has been evaded for far too long and, even when it is imposed its almost always based on the concept

[71] In his Shirley Williams Memorial Lecture, 15th October 2021 https://www.davidputtnam.com/viewNews/n/lord-puttnam-retirement-full-speech/

of 'fines'. Fines have never been enough. If we are serious about grabbing the attention of Boards towards their 'duty of care' then we have to create far clearer lines of accountability. 'Their instinct is to believe that they can always ride it out'.

If I could save only one element of the documented Business Architecture it would be the Principles, which represent the very soul of the Enterprise. They are so important, I want to dwell on them for a moment, and share a couple of pertinent examples. Firstly the Principles of Public Life, first set out by Lord Nolan in 1995, and applicable to all public servants (including Members of Parliament) in the UK. Apparently.

The Seven Principles of Public life

(The ethical standards those working in the UK public sector are expected to adhere to
Otherwise known as the 'Nolan principles'
They apply to anyone who works as a public office holder including:

- Those elected or appointed to public office, nationally or locally,
- Those appointed to work in the civil service, local government, the police, courts and probation services, Non Departmental Public Bodies, and in the health, education, social and care services, and
- Those in the private sector delivering public services.)

1. **Selflessness** – Holders of public office should act solely in terms of the public interest.
2. **Integrity** – Holders of public office must avoid placing themselves under any obligation to people or organisations that might try inappropriately to influence them in their work. They should not act or take decisions to gain financial or other material benefits for themselves, their family, or their friends. They must declare and resolve any interests and relationships.
3. **Objectivity** – Holders of public office must act and take decisions impartially, fairly and on merit, using the best evidence and without discrimination or bias.
4. **Accountability** – Holders of public office are accountable to the public for their decisions and actions and must submit themselves to the scrutiny necessary to ensure this.
5. **Openness** – Holders of public office should act and take decisions in an open and transparent manner. Information should not be withheld from the public unless there are clear and lawful reasons for so doing.
6. **Honesty** – Holders of public office should be truthful
7. **Leadership** – Holders of public office should exhibit these principles in their own behaviour. They should actively promote and robustly support the principles and be willing to challenge poor behaviour wherever it occurs.

And here are the principles on which the British police service was established almost two hundred years ago and which are as fresh and relevant today as when drafted by Sir Robert Peel

The Principles of Policing upon which the modern police service in the UK was founded in 1829

(Developed by Sir Robert Peel, and often known as the 'Peelian Principles')

1. To prevent crime and disorder, as an alternative to their repression by military force and severity of legal punishment.
2. To recognise always that the power of the police to fulfil their functions and duties is dependent on public approval of their existence, actions and behaviour, and on their ability to secure and maintain public respect.
3. To recognise always that to secure and maintain the respect and approval of the public means also the securing of the willing co-operation of the public in the task of securing observance of laws.
4. To recognise always that the extent to which the co-operation of the public can be secured diminishes proportionately the necessity of the use of physical force and compulsion for achieving police objectives.
5. To seek and preserve public favour, not by pandering to public opinion, but by constantly demonstrating absolutely impartial service to law, in complete independence of policy, and without regard to the justice or injustice of the substance of individual laws, by ready offering of individual service and friendship to all members of the public without regard to their wealth or social standing, by ready exercise of courtesy and friendly good humour, and by ready offering of individual sacrifice in protecting and preserving life.
6. To use physical force only when the exercise of persuasion, advice and warning is found to be insufficient to obtain public co-operation to an extent necessary to secure observance of law or to restore order, and to use only the minimum degree of physical force which is necessary on any particular occasion for achieving a police objective.
7. To maintain at all times a relationship with the public that gives reality to the historic tradition that the police are the public and that the public are the police, the police being only members of the public who are paid to give full-time attention to duties which are incumbent on every citizen in the interests of community welfare and existence.
8. To recognise always the need for strict adherence to police-executive functions, and to refrain from even seeming to usurp the powers of the judiciary of avenging individuals or the State, and of authoritatively judging guilt and punishing the guilty.
9. To recognise always that the test of police efficiency is the absence of crime and disorder, and not the visible evidence of police action in dealing with them.

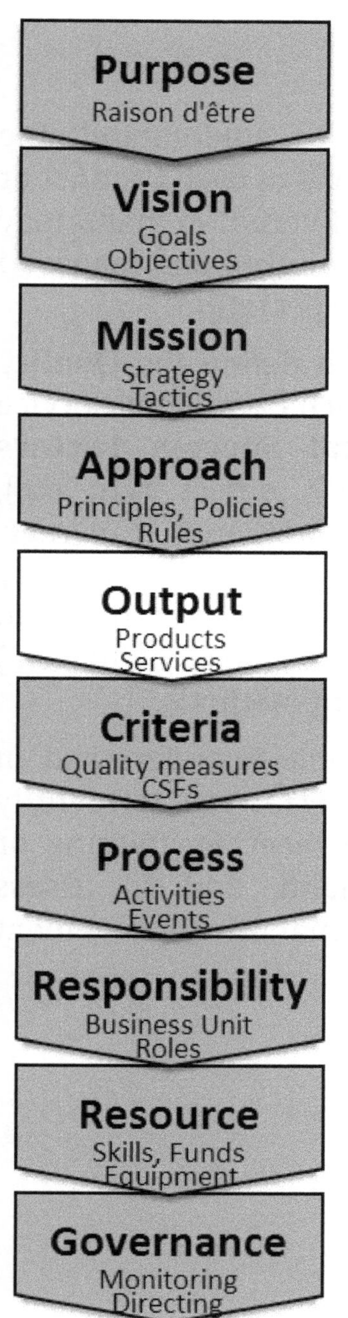

Outputs

The **Products** and **Services** delivered to customers or stakeholders with the aim of realising the Mission of the Enterprise.

They are where value is exchanged – typically products and/or services for money or funding, so are very important, and usually offer easy measures of performance and the basis for incentives.

Each should have defined business owners responsible for them and should be capable of expression in service contract terms, including performance measurements relating to those attributes which are most important (Key Performance Indicators), such as minimum and maximum volumes & quantities, availability timescales, quality standards.

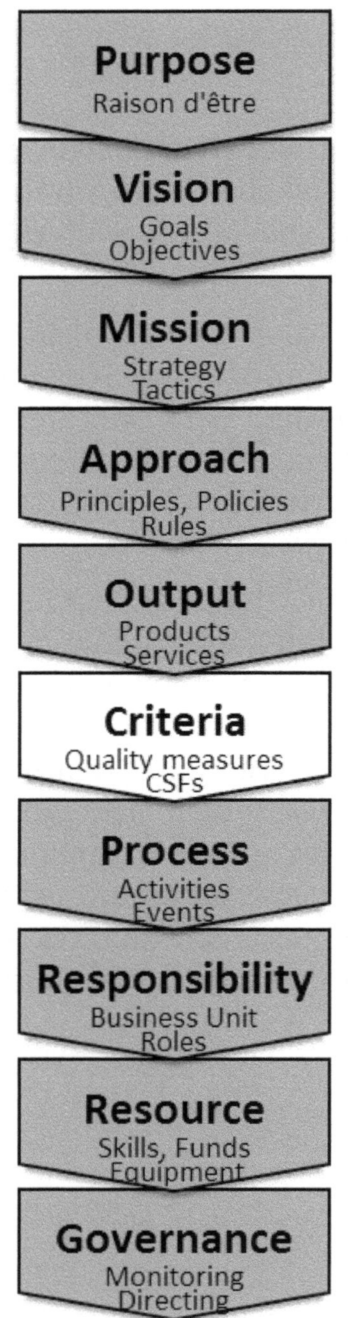

Criteria

It's important to understand the required attributes of each Product or Service being delivered, so that it may be designed, monitored and managed by that quality expectation.

Attributes should define the **Quality Measures** for each Product or Service and the **Critical Success Factors (CSFs)** (accuracy, defect rate etc), together with volumes (average, peak, minimum, maximum), delivery locations, timescales (order to delivery, frequency/seasonality etc) and measurement methods.

You need to be able to understand, in these terms, expectations, to identify whether you're meeting, missing or exceeding demand, and to assess whether you're making improvement or not.

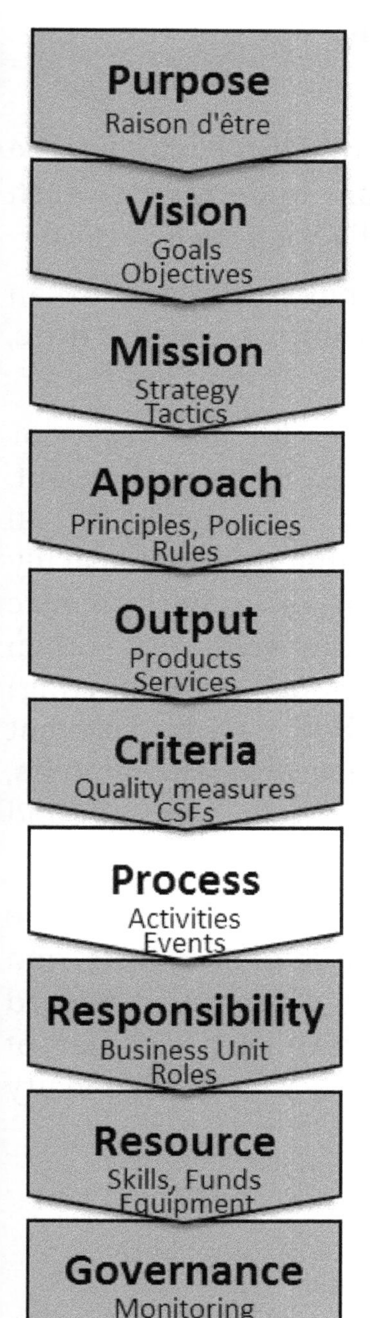

Processes

Processes are what the Enterprise actually does - the transformations it effects - and should be described in simple noun-verb terms. Hopefully you can see, by now, the importance of an Enterprise Glossary?

Crucially, processes are where value is generated.

It is also where resources are consumed, rules are given effect and responsibilities are fulfilled, so pretty central.

Processes are comprised of **Activities** and often respond to, or result in **Events**, which can be significant.

Each Process should have a designated owner and performance measurements relating to those attributes which are most important (Key Performance Indicator), e.g.s speed/duration parameters, accuracy/quality.

Events are typically when the state of something changes or is changed.

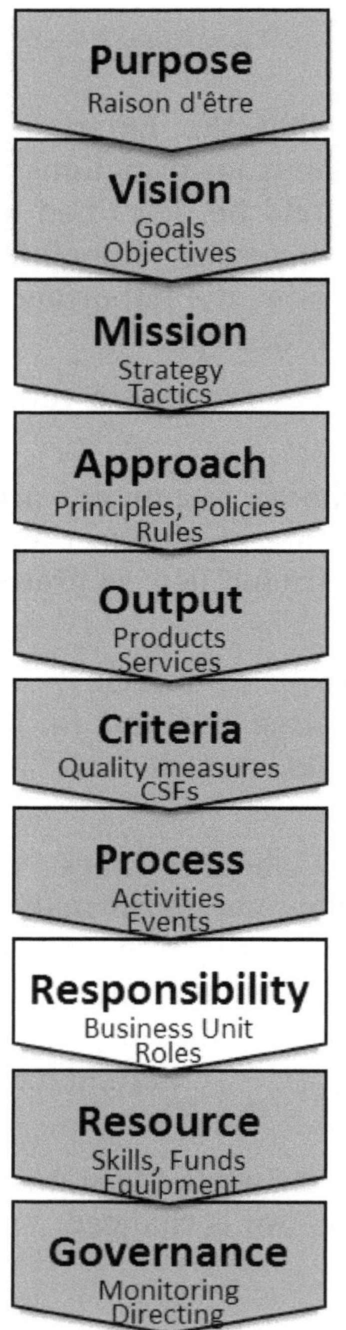

Responsibilities

Next you need to define who is to be responsible for undertaking each Process and delivering each Service.

Responsibility may be assigned to an **organisational unit,** a specific **Role**, or both.

The consequences for good or poor performance need to be spelled out. If responsibility is assigned to an organisational unit this will usually be in the form of a performance target, if to a specific role then through personal (reporting) objectives – both with relevant criteria and measurement metrics, and with associated bonuses and (actual or effective) penalties.

Note that a **Job** is comprised of one or (usually more than one) **Role**, and may, therefore, have a reward package comprised of a number of role-related reward/penalty elements.

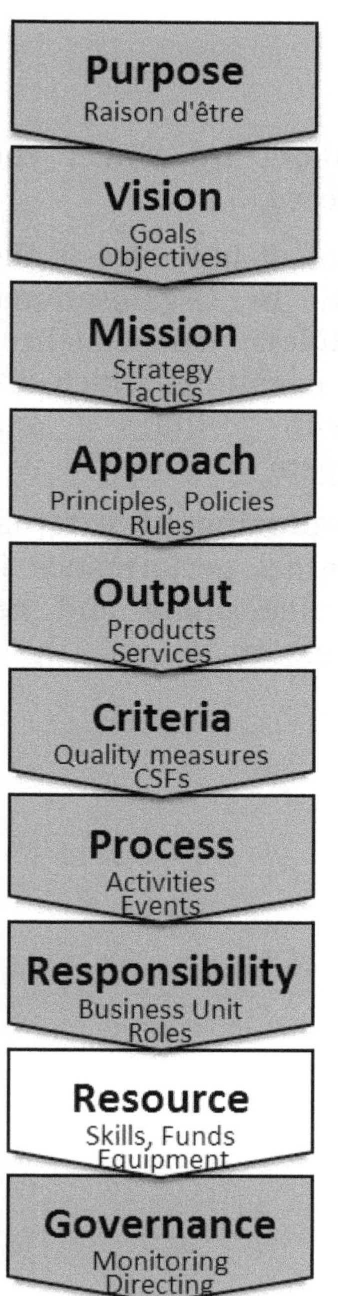

Resources

Resources are those 'corporate assets' which may be utilised or consumed during the performance of processes.

Corporate assets include funds, knowledge (including IP[72]) and authority.

Remember that although corporations are fond of saying that their greatest assets are their people, and it might be worth recognising the value that they bring, employees don't belong to them.

[72] Intellectual Property

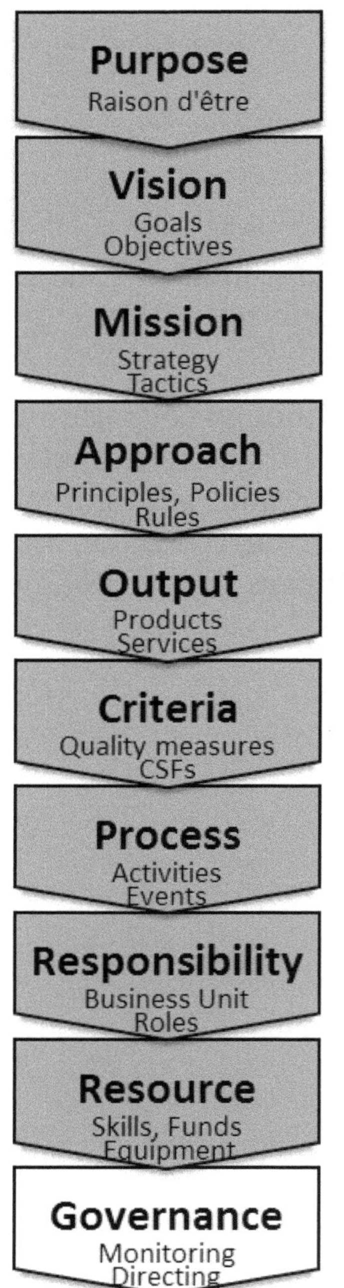

Governance

Finally, the "tiller" by which the Enterprise is steered.

The primary root of influence is the ability to apply to Organisational Units and Roles, consequences (incentives and penalties) which are associated with Rules and performance targets.

Continuous **monitoring** of external change and internal performance is required, with fine-tuning and re-**directing** as necessary.

Having done all of the above you will have a pretty good view of your Enterprise, and foundation for much else.

Only a few need to be exposed to the public, but in my many years of experience I've seen few good examples of things which are described as them in major Enterprises; plenty of 'missions', aims and 'principles'/'values' which turn out to be little more than marketing straplines.

I was impressed by the plaque in a hospital reception a few years ago, proudly displaying theirs[73], which was a notable exception. It said all of the things I wanted (and expected) to see and gave me confidence that they had probably used as much rigor on the other levels.

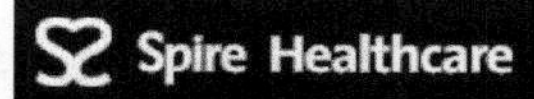

Our Vision

To be recognised as a world class healthcare business.

Our Mission

To bring together the best people who are dedicated to developing excellent clinical environments and delivering the highest quality patient care.

Our values

We are extremely proud of our heritage in private healthcare and of our values as an organisation:

- Driving clinical excellence
- Doing the right thing
- Caring is our passion
- Keeping it simple
- Delivering on our promises
- Succeeding and celebrating together

Our people are our difference, it's their dedication, warmth and pursuit of excellence that sets Spire Healthcare apart.

Compliance

We have a strong culture of compliance, which applies at all levels of the business and is supported by our compliance programme. Spire considers that it is the responsibility of all Board Members, staff and representatives of Spire, at all levels, to ensure compliance with all applicable laws, including competition law, anti-bribery law, anti-tax evasion facilitation law, healthcare regulations and data protection law.

The Board is determined that strict compliance with the law is a requirement of the business. No director or employee has the authority to give any order or direction which conflicts with this.

Next, you need something called a 'Capability Model'

---

[73] Spire Healthcare Group plc. This version taken (with permission) from their web site.

Capability Model

A Capability Model is a very important element of your Business Architecture. It's just a fancy way of saying that you need to understand what your Enterprise needs to be able to do.

The dictionary definition of capability is 'the quality of being capable; ability' but, in this context I prefer to use the one from DODAF[74] – 'the ability to achieve a desired effect under specified standards and conditions through combinations of means and ways to perform a set of tasks.' In my, technology agnostic, view, I prefer the concentration on the achievement of an effect, rather than how it is achieved.

The Capability Model should:

- Represent only Activity
  Not responsibility, volumes, sequence, location, periodicity or supporting resources
- Be "implementation-neutral" (and hence stable)
- Cover the full range of the Enterprise's activities
  Without duplications or exceptions
- Use the language of the Enterprise
  Plain English and generally understood terms
- Be expressed in simple verb-noun terms
  e.g. 'Investigate Crimes'
- Each activity should produce a valued product, output or outcome

---

[74] The Department of Defense Architecture Framework (DoDAF) is an architecture framework for the United States Department of Defense (DoD)

ie they should be “value-oriented”

It doesn’t have to be right. Arguably it can never be right (this is a good example of the observation that ‘perfection is the enemy of the good’[75]). But it needs to be right-enough; something which everyone in the Enterprise can recognise, rally behind, and move forward. That can be difficult, and take a lot of work. But it is necessary and worthwhile.

Adding short descriptions to the capabilities helps to clarify what it is, and to aid stakeholders’ thinking process. Those short descriptions should use plain business language, make clear what is and isn’t included, and be oriented to describing elements in terms of what each enables the organisation to do.

Here is the one which we came up with for the Police service. With 43 autonomous forces, it took heaps of internal consultation, lots of patience and perseverance, and a long time.

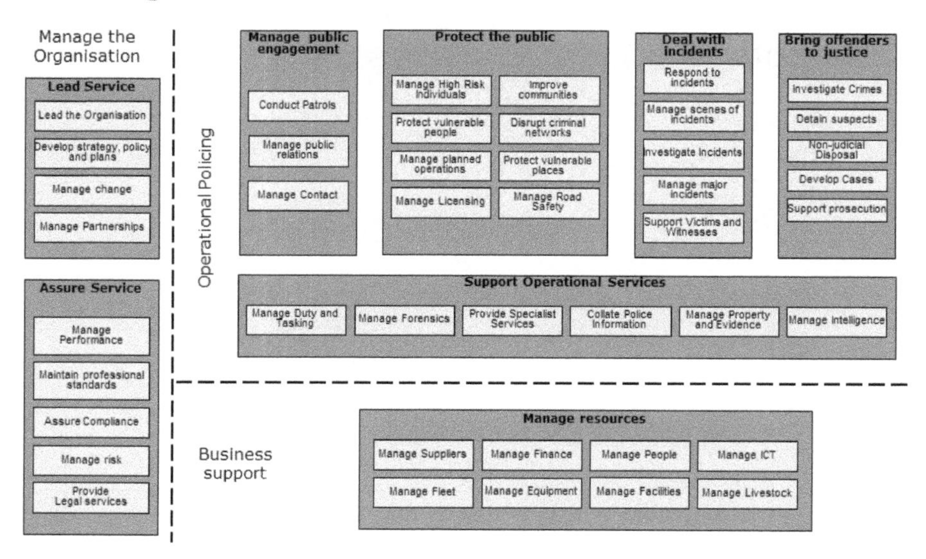

[75] Voltaire

I would expect every Enterprise to include those three core segments: Managing & directing, Core operations, and Support functions.

Models can be broken down one or two levels to whatever is useful or most meaningful to stakeholders. Doing so may make the descriptions unnecessary, and may result in duplications (which need to be rationalised) at low levels of granularity.

You'll more often see IT Capability models, showing IT modules or components. These ought to (but often don't) begin at Business Capabilities, and demonstrate how IT Capabilities support them. I created this one to illustrate explicitly how, in a modular, component architecture, strategic IT systems might be supplemented by tactical components – many of which would be usable to support a variety of Business Capabilities. The aim was an alternative to endless development of large, monolithic systems, and the creation of a marketplace for reusable commercial components.

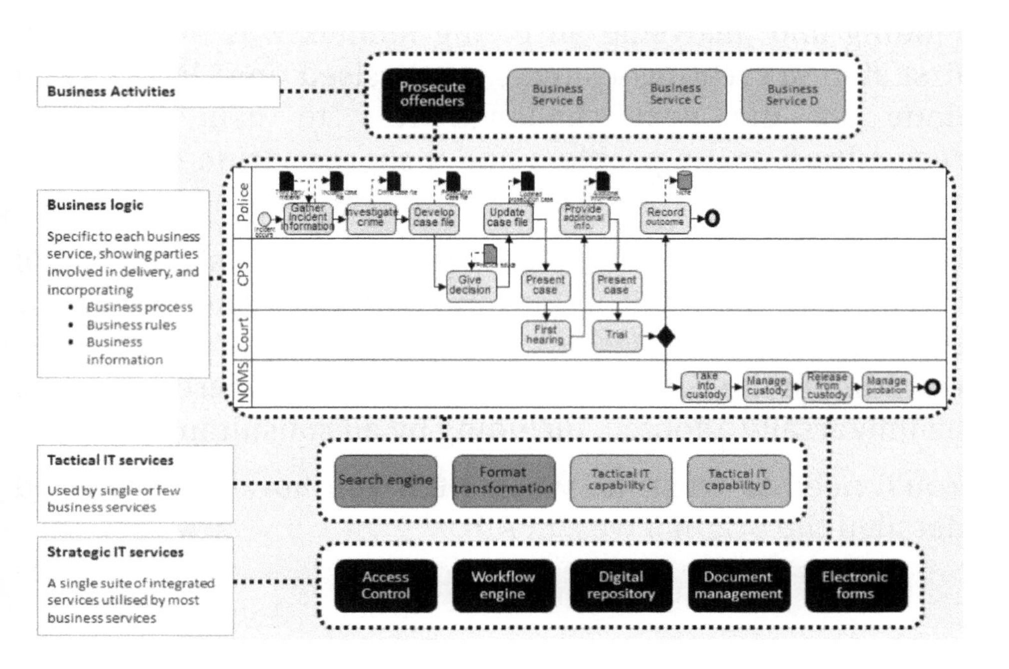

Later I'll talk about how important the Business Capability model is and how it can be used, but creating it and getting "buy-in" can be hard work. If you're lucky the need will already have been identified and a fully funded project to create it established. The real world is rarely so kind.

In UK Policing there were already more than 20 competing models of what the service did. Most Chief Constables and many departments and interested parties had developed their own, and seemed happy with them. The diversity resulted from fresh models being created to address each new need that arose, and that tactical approach appeared to have been extensively exploited by consultancies. But I could see that it was inefficient, unhelpful for collaborating agencies and cross-service suppliers, and hindered collaboration and integration between forces.

Finding and analysing all of the models was an essential first step in creating a single, synthesised one. When I had done so, the next challenge was to gain Executive recognition of the problem and a shared vision of a better approach.

I created an A0[76]-sized, colour "poster" which showed all of the models, pinned it in a prominent place and waited for the inevitable discussions. The result was spectacular, resulting in the commissioning of a new standard model, to be universally adopted, including by all consultants.

You'll need to trust me when I tell you how colourful and detailed the original was, in A0 .

[76] ISO 216 standard (841mm x 1188 mm). The A Series paper sizes cleverly halve each size, so that for example A4 is half of A3 size paper, which is half of A2 which is half of A1, which is half of A0, which itself is $1m^2$, which means that you only really need to remember one paper size in order to remember them all.

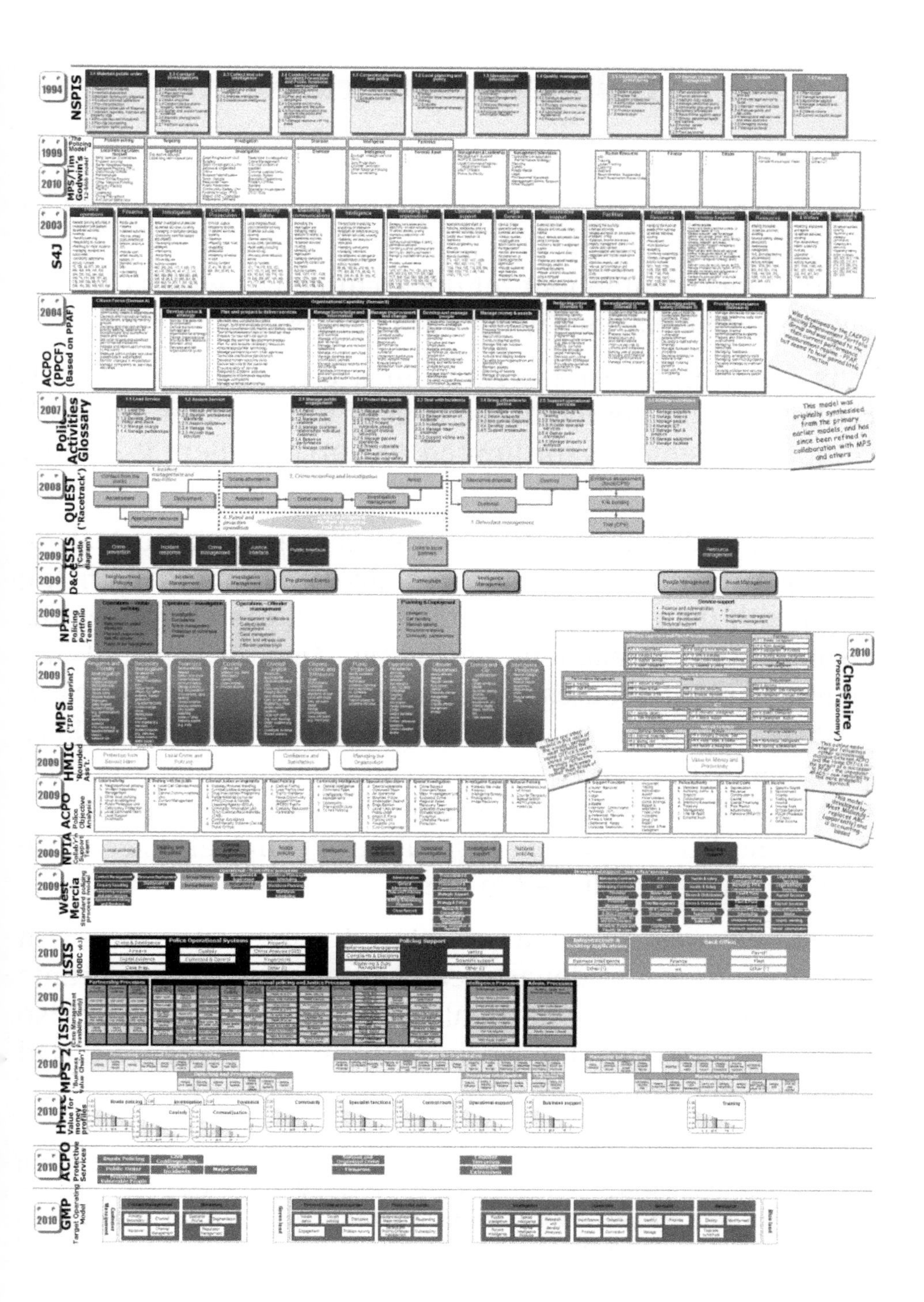
1994 NSPIS
1999 'The Policing Model'
2010 MPS/Tim Godwin's '12-blob model'
2003 S4J
2004 ACPO (PPCF) (Based on PPAF)
Was developed by the [ACPO] Policing Improvement Portfolio Group and was aligned to the then current performance measurement regime - PPAF, but seems to have gained little traction
2007 Police Activities Glossary
This model was originally synthesised from the primary earlier models, and has since been refined in collaboration with MPS and others
2008 QUEST ('Racetrack')
2009 ISIS ('Castle diagram')
2009 D&C
2009 NPIA Policing Portfolio Team
2009 MPS ('IPI Blueprint')
2010 Cheshire ('Process Taxonomy')
2009 HMIC 'Rounded Ass't.'
2009 ACPO Police Objective Analysis
This model - developed by West Midlands - replaces ABC (apparently) and is accounting-based
2009 NPIA Collab'n Support Team
2009 West Mercia Standard policing processes model
2010 ISIS (SOBC v0.3)
2010 ISIS (Crime Management Feasibility Study)
2010 MPS 2 ('Business Value Chain')
2010 HMIC Value for money profiles
2010 ACPO Protective Services
2010 GMP Target Operating Model

Balanced Scorecard[77]

A Balanced Scorecard is a consistent measure of worth to the Enterprise – what it values, in relative terms - and is another key "component" of the architecture which can and will be used in various ways.

It is something to be created in a calm, logical manner, rather than on the fly, in the heat of the moment and with competing demands for limited resources, where it is most valuable.

As an extreme example, consider the unenviable task of having to ration healthcare resources. Fraught with life and death decisions and a finite budget, the ethical rights, wrongs and thresholds, including 'Quality-adjusted life years'[78] are all debates which need to be had, but not case-by-case. It was to avoid a 'Postcode lottery', where treatments that were available depended upon the NHS Health Authority area in which the patient happened to live, that The National Institute for Clinical Excellence (NICE)[79] was established in 1999.

Its status was changed in 2013 from a special health authority to a non-departmental public body and became the National Institute for Health and Care Excellence. It also officially took on the work of bringing evidence-based guidance and standards to the social care sector.

---

[77] The original 'Balanced Scorecard' was developed by Robert Kaplan and David Norton in 1992 and incorporated four perspectives: 'Financial or Stewardship' (financial performance, resource use), 'Customer and Stakeholder' (customer value, satisfaction and/or retention), 'Internal Process' (efficiency, quality), and 'Organisational Capacity' or 'Learning and Growth' (human capital, infrastructure and technology, culture)

[78] Various definitions, but see NICE Glossary for theirs

[79] https://www.nice.org.uk/

Its role is to improve outcomes for people using the NHS and other public health and social care services. It attempts to assess the cost–effectiveness of potential expenditures within the NHS to determine whether or not they represent 'better value' for money than treatments that would be neglected if the expenditure took place.

I wouldn't want that responsibility, but I'm glad that someone is doing it in a measured, scientific and considered way.

For most Enterprises the challenge is probably less consequential, but nevertheless still vitally important. It supports

- Alignment of performance measures and KPIs with overall Enterprise goals;
- Setting of personal, departmental and organisational objectives;
- Better project prioritisation.

Every Enterprise should have its own, tailored scorecard, which needs to

- Accurately reflect the core purpose and values of the Enterprise;
- Incorporate all of the diverse products and services it delivers, in a way that enables side-by-side comparison;
- Remain constant over a long period of time (subject to variations only in detailed, specific measures or in weighting of relative importance);
- Be as simple and straightforward as possible;
- Support the various different purposes in a common, consistent manner;

The key point is that all change initiatives and operational activity can have broad effects, some of which might be less immediately apparent than others.

Below is a high level view of the one that I developed for the Police service in 2007.

At that time Despite the welter of data required by the Home Office in the ADR (Annual Data Return) what was missing from the policing landscape was a universally accepted definition of what constituted 'good' or 'bad' in terms of outcomes. In common with many Enterprises, judgements were made about the worth to the organisation of a variety of things using separately devised, disconnected sets of measures. Examples included local plans, individuals' personal objectives, project selection and prioritisation and performance measurement and management.

In Policing, the most authoritative (but, by no means only) measurement mechanism was called PPAF (Police Performance Assessment Framework – those with acute eyesight may have noticed the acronym amongst the many others on the monster 'capabilities model' poster a few pages back), although this too was driven by the Home Office to exercise an additional control over the police service by exposing apparent 'performance differentials' through the 'league tables' that are invariably published by the press.

Paradoxically, and tellingly, a separate set of measures was often used for performance management by individual forces at local level.

Operational outcome measures still predominate, but they are joined by others, many not recognised within PPAF.

Even when including financial return within an Enterprise Balanced Scorecard you need to recognise the different potential value-related measures. For examples, Economic Value Add (EVA), Return On Equity (ROE), Return On Assets (ROA), Return On Capital Employed (ROCE), Payback Period and so on. These will allow you to accommodate different perspectives and strategic priorities over time. For example, is the priority a quick but modest return, bigger returns in the longer term or even a defined loss as a price to pay for building market share?

The next step is to add weightings, which can be adjusted over planning cycles, to reflect changing priorities. But within each planning cycle, the same weighted factors will be applied equally to every proposal, to ensure objectivity and to enable comparison over time and the identification of trends.

Rules

Rules define acceptable or unacceptable states of affairs that may result from a Course of Action. In other words, they are intended to govern Activity.

The Principles on which they are founded are usually enduring characteristics of the Enterprise.

The OED defines a rule as a 'principle to which action or procedure conforms or is bound or intended to conform'. From which it's easy to understand why they are so important in Engineering Systems, which are designed to be compliant.

But, as I said earlier, people do not follow rules. At best they pay attention to consequences, which may or may not be associated to rules, and to incentives.

It's important to understand how this works, and a model can be helpful. Here's the one I created for policing.

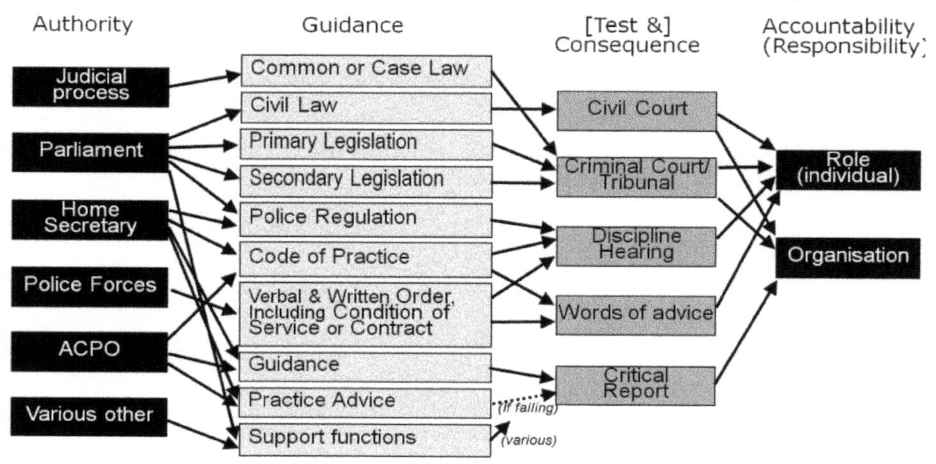

It's also often useful to understand how influence and 'soft power' can be exerted.

Here's an example model that I created in about 2005 - again for policing

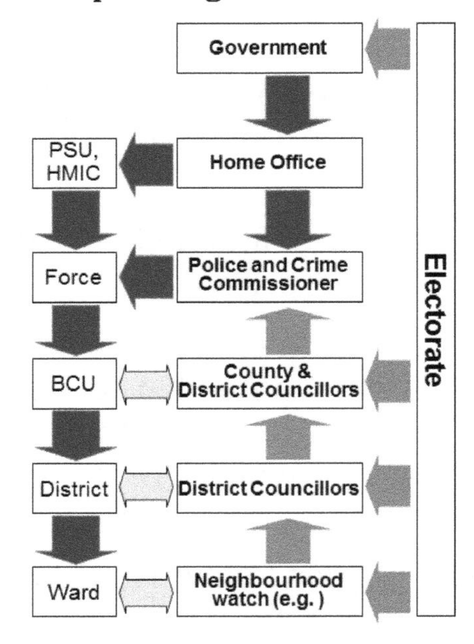

Governance and Influence in policing

Governance Depends on the ability to apply consequences (and thereby invoke motivators)

'Influence' reflects the ability to indirectly affect the application of consequence, via governors

Key:

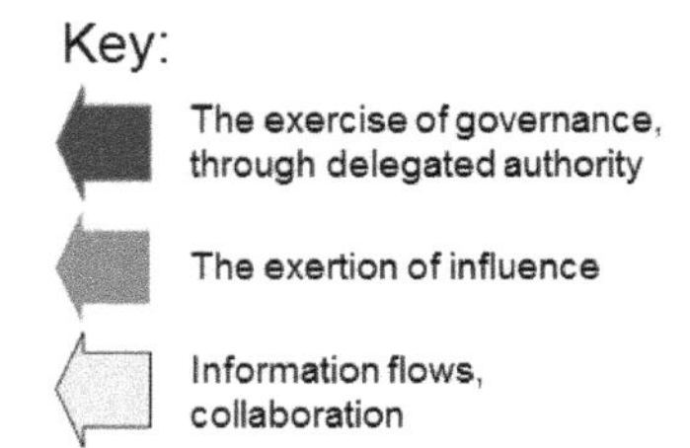

PSU = Police Standards Unit
HMIC = Her Majesty's Inspectorate of Constabulary
BCU = Basic Command Unit
District and Ward are just geographically-aligned organisational units

Or to understand funding and accountability relationships, as with the changes made to public health provision in 2013

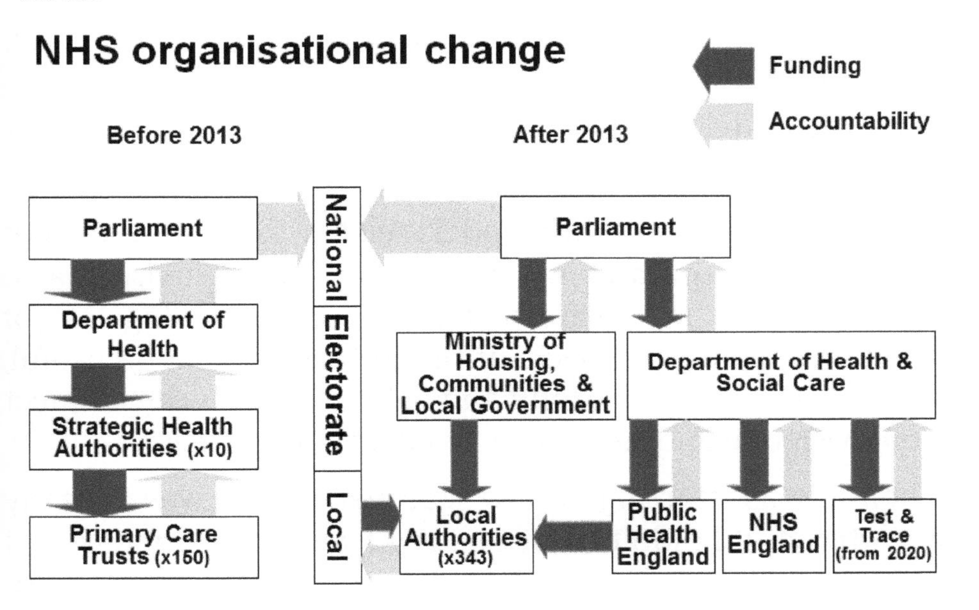

Note that the 'Test and Trace' function was set up separately in 2020, rather than utilise existing capabilities within the NHS.

Metamodel

Although there are various definitions, a metamodel is generally understood to mean a model of models; the rules governing the creation of other models.

My use of the term 'Metamodel' relates to a guide on not only how the separate models need to be designed, but also showing how they relate to one another. Having lots of models of separate aspects of the Enterprise can be useful, but they are far more useful and valuable if you can connect them with one another; that's what architecture is.

When I was starting out I searched what was already available, and I didn't want to stray from open standards. I surely wasn't the first to grapple with this? It turned out that amongst the few interested parties there was nothing which fitted the bill.

The most promising source was the OMG[80]'s 'Business Motivation Model' which was then being developed[81]. Its attraction was that it offered the business, rather than an IT perspective. It promised to

- Identify factors that motivate the establishing of business plans;
- Identify and define the elements of business plans;
- Indicate how all these factors and elements inter-relate.

---

[80] OMG – the Object Management Group – see earlier footnote (#29)

[81] The model was originally developed by Initially developed by the Business Rules Group (BRG), and was adopted by the OMG in 2005. V.1 was released in 2008 https://www.omg.org/spec/BMM/1.0/PDF, and the latest version, 1.3 https://www.omg.org/spec/BMM/1.3/PDF, was released in 2015

This is an overview of the model (version 1.0 through to version 1.3)

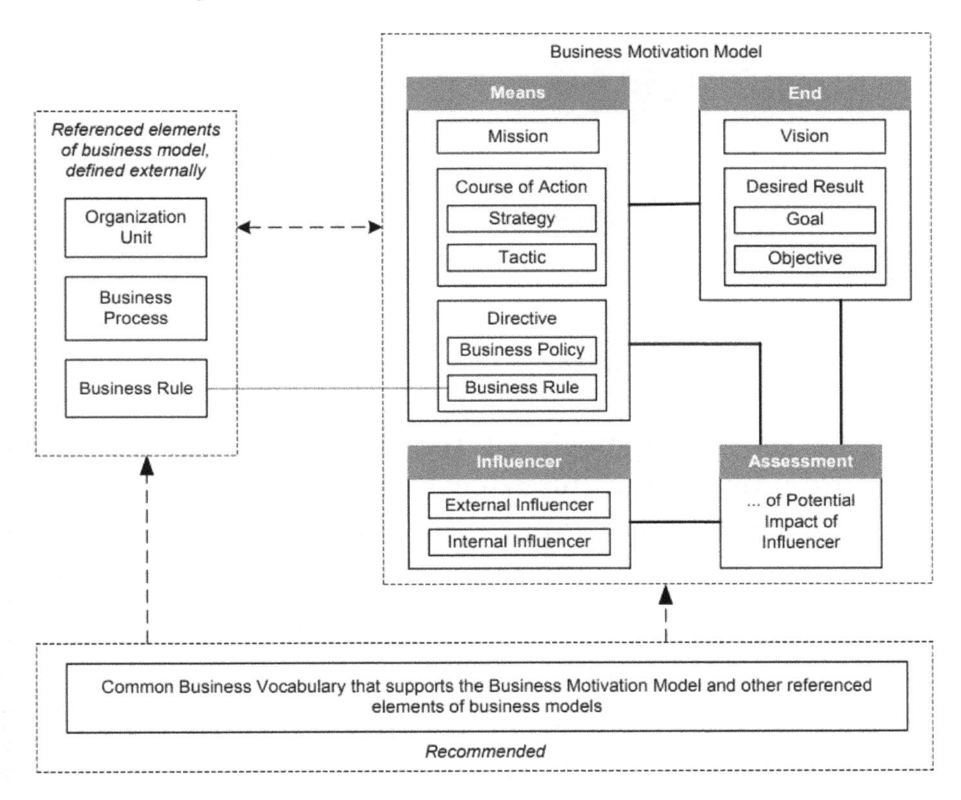

It was a good start, but not nearly enough for me - I was under time pressure to deliver a workable definition of the business Architecture, and I needed to cover a much wider scope[82].

After much additional work, numerous iterations and refinements, here is the version which I came up with. It is generic and can be applied to any Enterprise. I call it a Concepts Metamodel.

[82] The OMG acknowledges in the specification that the BMM does not offer a complete business model framework.

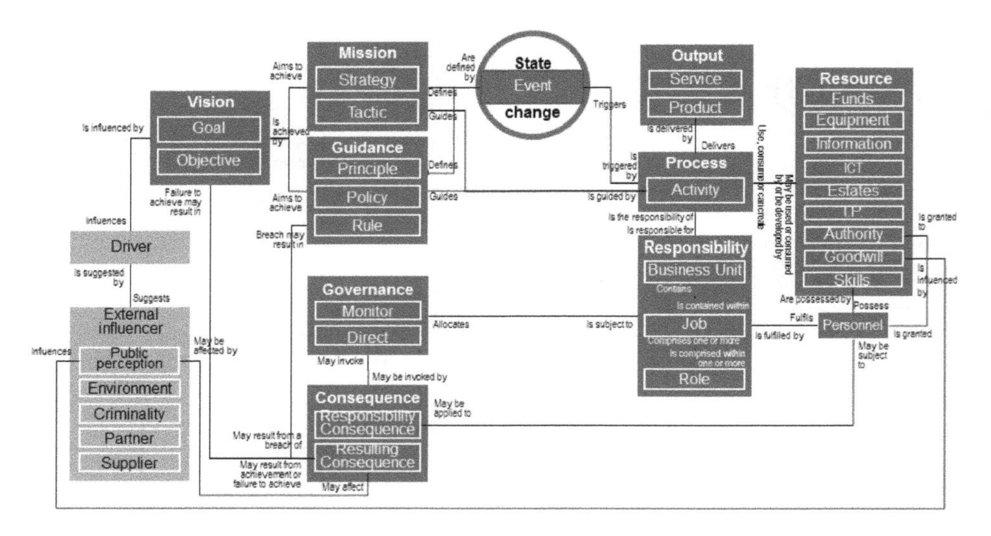

You can see the genesis of the model in the 'Business Motivation Model and the areas that I needed to "build out" – notably to the right where the real action takes place - Processes. I've also defined the nature of the connections.

I'll just say a little more about the 'Event' element ('state change'), because that is at the heart of the dynamics.

A Business Event is a point in time at which something significant occurs, such as something new coming into existence, or something changing. It can be either an external event which triggers a process, or a change in the state of something as a result of completing another process.

Typically this is the point which "governs" initiation of a process, or activity. It reflects the fact that processes are usually triggered by certain pre-set conditions having been met, just as a thermostat switches on or off your central heating in response to a change in ambient temperature. This allows the conditions that trigger the process or activity and what the process is to be separately defined.

The following diagram aims to explain that concept better.

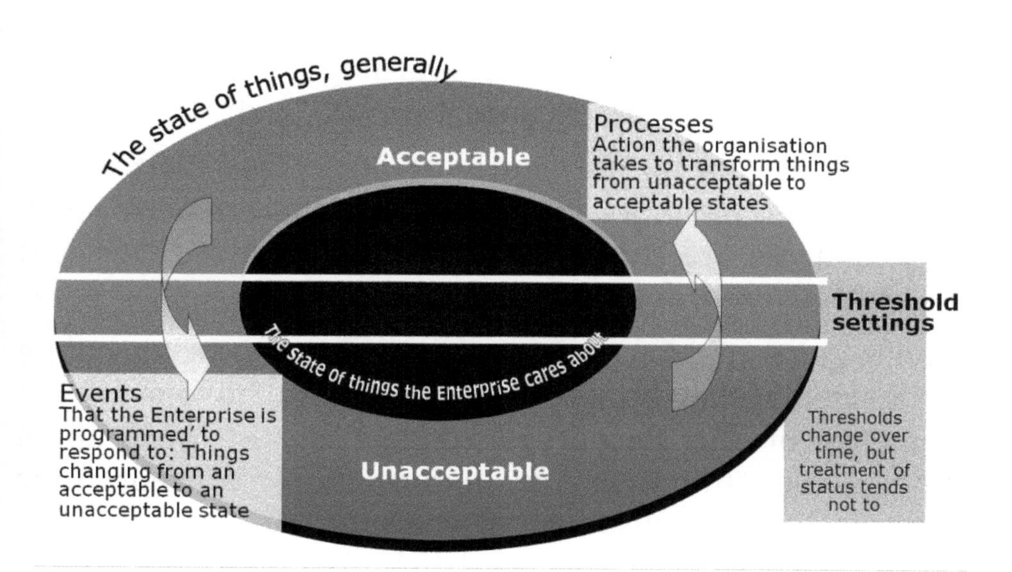

This was a useful model for me in relation to UK law enforcement. In the case of cannabis for example, you could have consistent processes for dealing with possessors for personal use, and for dealing. Thresholds would be legislation-based rules, which could be separately set and managed (The rules and thresholds in relation to cannabis have changed several times in the UK in recent years[83]).

A Business Process is a series of Activities that add value by transforming things from one state to another and to provide a business response to an initiating event.

Business Processes are pivotal for two reasons:

a) They generate value for the Enterprise (if you think they aren't, stop doing them. But be careful before you

[83] In the UK until 2004 Cannabis was classified, under the Misuse of Drugs Act 1971, as a 'class B' drug, when it was re-classified as 'class C', with differential penalties depending on whether it was for possession [personal use] or 'intent to supply' (with associated quantitative measures). In 2009 its classification was returned to 'C', but it has been a continual source of debate since.

do, remembering my earlier warning that things tend to be the way they are because it works), and;

b) It is in the performance of Processes that many elements converge. For examples, Roles are performed, Resources are consumed, Rules are followed, Services are delivered, and Products are created

Having generated value, that value ought to show up and be recognised somewhere. It should accrue to one of the elements of the Resources concept in the Metamodel (under the 'Goodwill' or 'Information' concepts, for examples), the realisation of Goals or the creation of a Service capability. The yardstick for measurement of 'value' is, of course, the Balanced Scorecard.

Someone – an organisation and a Role – should be an assigned owner of each and every Process. Remember that each Job comprises one or more Roles.

These basic components are useful in isolation, and enable consistent recording, referencing and linking, but combined they offer a powerful strategic capability.

Putting the Capability model to work

Too often, in my humble opinion, 'Business Architecture' is used as a means to justify strategic decisions which have already been taken - sometimes even to create links to technology solutions which have already been selected – but it is capable of helping to identify or prioritise the need for improvement and to be the driver of change.

The following diagram shows the capability model at the centre of a "Heat Map", where the demand/requirement for each capability is contrasted with current capacity to deliver it.

At this stage you can think of capabilities and services interchangeably, and remember that you can produce capability models at lower levels of abstraction – whatever level is needed to provide useful insight. Sometimes it can be useful to consider capabilities at a relatively small, granular level and, as you'll see later, it is often vital to think about services required in terms of their wider context in order to better understand critical dependencies and overall performance standards required.

You need to define each capability in detailed terms including, but not exhaustively:

- Who does it deliver to?
- What medium does it need to be delivered via (physical/ face-to-face/ on-line/ via telephone/ other)?
- What are the required hours of operation/availability?
- Where does it need to be delivered (geography)?
- What volumes/frequency/ capacity are required?

- What quality standards are required?
- What are the critical dependencies and performance characteristics?
- How do your delivery capabilities compare with those of others in the market?

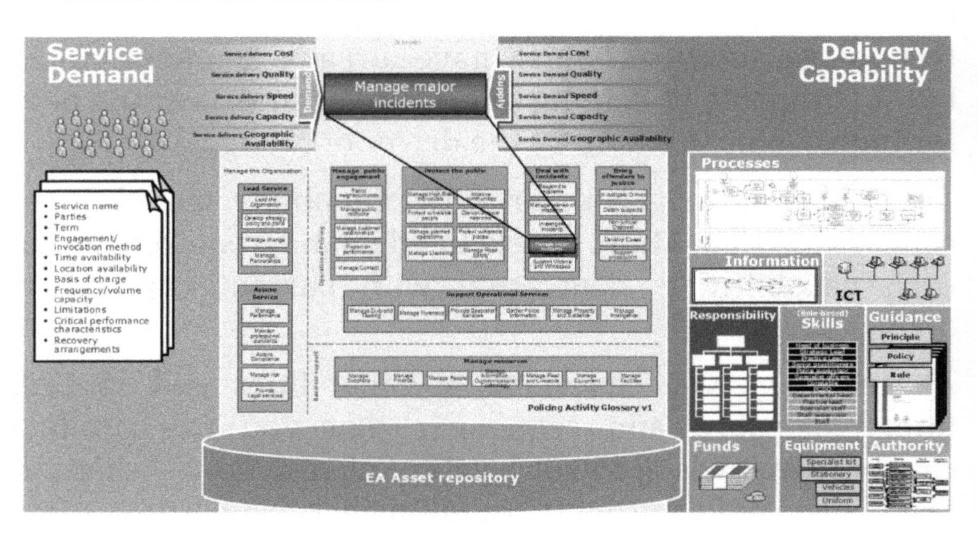

An example Capability 'Heat Map' based on the police service

Now is the time to think about whether your Enterprise has unique or market-leading capabilities - perhaps one or a few represent your USP[84] and need to be cherished. Also to think about whether a capability which you have has a wider market - remember the translation services of Xerox.

One option for fulfilling each required capability will be to sub-contract or buy it in as a service from a third party, so it is vital to be rigorous with defining terms, since these could form the basis of a commercial contract. If you do take this course, you must carefully consider the fact that whoever fulfils the capability for you will not share your

[84] Unique Selling Proposition

Enterprise Purpose, Vision, Mission or Goals - each Enterprise has its own. If it is a Limited company then it will have a legal obligation to act in the best interests of its shareholders, for example, not yours.

You would need to draw the terms of the contract very tightly, to ensure that you are comfortable with the level of residual risk, in terms of profitability, business continuity, workflow and reputation. If you can't, even with penalty clauses and robust commercial guarantees, then it's probably not a suitable candidate for 'outsourcing'.

Once completed you should have an understanding of all of the Capabilities that the Enterprise needs and any under- or over- provision. A colour-coded version of the Capability Heat Map can then form a useful graphic to draw attention to service changes or "tweaks" needed, which can be defined, prioritised, and used as the foundation for planning.

Different perspectives

To check for completeness – to make sure you've captured everything – it's often useful to look at the Enterprise from unfamiliar perspectives. The tendency is to take an insider's view, but what about from outside? From a customer's, a supplier's or a competitor's point of view, for examples?

When I led a team of Business Analysts, one of our key responsibilities was to gather and document user requirements as the basis for developing software. We used a modified version of a standard method, called RUP[85], and UML[86] for doing this. One key element is a 'Use Case', which is defined as *'a description of a set of sequences of actions, including variants, which a system performs to yield an observable result of value to an Actor'*. 'Actors' are people or systems that engage with the system (in order to gain that value). Each Use Case describes the 'main flow of events' (often called the 'happy day scenario') and one or more 'exceptional flow of events' (to cover what happens when things go wrong – like entering an incorrect PIN in an ATM or card payment terminal). Our practise was to go a stage further and model, additionally, a "malevolent actor" - someone who was deliberately trying to exploit, circumvent or otherwise misuse the system. We found this

[85] The Rational Unified Process. Rational Software has been a Division of IBM since 2003.

[86] The Unified Modeling Language – an open standard which was originally developed by Rational Software and which, in 1997, was adopted as a standard by the Object Management Group (OMG).

practice especially useful in an arena where some parties were trying to do just that.

It led me to consider, in relation to policing, something in a similar vein, but on a much broader scale - a model of the criminal enterprise, or at least of economic crime; the "crime economy". My paymasters were content with just modelling the policing business architecture, and weren't prepared to fund this work, so it didn't happen. As a taxpayer you might agree, but I remain convinced of its potential value.

As with all Enterprises, elements of crime are well known, but rarely an overview of the whole "economy". Lots of crime is, of course of a very trivial nature – pilfering of office stationery, littering, for examples – but it's an open secret that not all luxury lifestyle trappings are bought with honestly earned cash, and not everybody enjoys a safe life. When it comes to economic crime there is a whole parallel economy, ranging from tax-evasion, through the sale of illegal drugs, fly-tipping, car theft, forgery, blackmail, car 'clocking', shoplifting, fraud, the sale of stolen or counterfeit goods, 'mugging', and identity theft, to armed bank robbery. The cost is borne by us all and the world is full of people who, for various reasons, can and do make the lives of the rest of us a misery.

A detailed analysis would require a separate, book, but the basic premise that I had in mind was to treat the world of crime as an Enterprise in itself, and to model it like any other. Starting with the overall objective of creating and sustaining illegal activities, the scope would, of course, need to be defined and agreed, but we know that that parallel economy can be regarded as analogous to the 'malevolent actor', and represents a livelihood for a portion of society. Much is what occupies police time - law enforcement and criminal justice, anti-social behaviour and public safety and protection. A recent development in their tactics relating to drug trafficking has been to switch from trying to track the drugs to tracking the proceeds of crime - money.

When restricted, illegal activity tends to migrate to another, often similar, activity, or switches are simply made from one to another with a better risk/benefit profile. For example, when enforcement capabilities were enhanced and penalties for drug smuggling were increased many smugglers switched to smuggling alcohol and tobacco, from which good incomes could also be made, and for which penalties were far less harsh.

Another justification for developing a better, more structured understanding of the world which law enforcement is engaged to address is the deleterious effects that it has. Other public agencies, such as those dealing directly with victims (including people who are exploited) also expend a great deal of effort dealing with the consequences of economic crime. Then there is lost tax revenue, the cost to the NHS, insurance loss and so on.

I never understood why, for example, we all appear to be relaxed about the fact that there are around 1 million uninsured cars on Britain's roads. That is likely to be £600m - £700m[87] pa extra that the rest of us have to pay on the 34m vehicles that ARE insured. There are, of course, other costs. In the UK each year 130 people are killed and

---

[87] Money Helper – from Statista 2020 data https://www.moneyhelper.org.uk/en/blog/car-insurance/what-is-the-average-cost-of-car-insurance

26,000 suffer injury after being involved in a collision with an uninsured or untraced driver. This equates to one casualty every 20 minutes. Additionally one in every five collisions on the country's roads involves an untraced or uninsured driver[88]. A total of 100,983 drivers were caught driving without insurance in 2020 and 102,387 vehicles seized from offenders[89].

Autocar reported[90] on 30th November 2021 that 'An estimated 1.9% of vehicles on the road in the UK are unlicensed and avoiding vehicle excise duty (VED)'. 'The DfT estimates the owners of 719,000 vehicles in active stock haven't paid the tax, with a potential revenue loss of around £119 million over one year'. The abolition, in 2014, of the paper tax disc is probably a key factor.

All of which ought to provide a sizeable pot – the thick end of a £Billion - to fund efforts to address the issue. That could involve data analysis (insurance details are provided by insurers to the police, who have alert facilities on ANPR to identify from both fixed and in-car cameras uninsured or untaxed vehicles on the road), followed by a programme of door-knocking and street stopping. For context, the whole of UK policing costs around £16Bn/yr[91]

---

[88] Kent police news, 15th November 2021 https://www.kent.police.uk/news/kent/latest/policing-news/national-campaign-targets-uninsured-drivers/

[89] National World https://www.nationalworld.com/lifestyle/cars/police-crush-56000-cars-as-uninsured-driving-soars-3385075

[90] https://www.autocar.co.uk/car-news/business-government-and-legislation/new-uk-government-figures-reveal-one-50-cars-untaxed

[91] https://www.gov.uk/government/statistics/police-funding-for-england-and-wales-2015-to-2022/police-funding-for-england-and-wales-2015-to-2022

Development of an enterprise is often driven by market competition, in much the same way as Darwin realised that evolution of the animal kingdom is and has been shaped by competition over millennia. Particularly in fields where there is market competition for discretionary spend, enterprises need to be nimble and agile or will likely perish. Smart ones will incorporate into their corporate fabric mechanisms to encourage "mutations" - internal competition and testing of ideas - before their competitors do. Formal examples include Lean[92], Six Sigma[93], Kaizen[94], and staff suggestion schemes.

Earlier, when discussing the Purpose of the Enterprise, I spoke, from my experience, about Enterprises engaged in the production and/or sale of products and/or services, and later about how the Capability Map can be used to consider opportunities to outsource elements. In more recent years, opportunities have been identified to improve products or services by adding to their scope to enhance the overall offering. Sometimes the uplift in value is so great that the new party can become perceived as the lead provider, relegating the original one to a support role. By 2015 Tom Goodwin noted[95] that ***'Uber, the world's largest taxi company, owns no vehicles. Facebook, the world's most popular media owner, creates no content. Alibaba, the most valuable retailer, has no inventory. And Airbnb, the world's largest accommodation provider,***

---

[92] Lean manufacturing is a production method aimed at reducing times within the production system. It is closely related to just-in-time manufacturing https://en.wikipedia.org/wiki/Lean_manufacturing

[93]a set of techniques and tools for the continuous improvement of processes to analyze and reduce errors or defects https://en.wikipedia.org/wiki/Six_Sigma

[94] Process of Japanese origin of implementing continuous improvement https://en.wikipedia.org/wiki/Kaizen

[95] in 'Digital Darwinism: Survival of the Fittest in the Age of Business Disruption'

owns no real estate. Something interesting is happening.'

Tom could have added other dominant players like Amazon, eBay Paypal, YouTube and Autotrader, and since he wrote it, numerous suppliers in various fields, such as Rightmove, Just Eat, Go Compare, Cazoo and HelloFresh seem to spring up at an alarming and accelerating rate. Many will, doubtless, fall by the wayside. Some result from the availability of new technology or changes in social trends (the COVID pandemic was undoubtedly a significant catalyst for many), and often involve a race for 'first mover advantage', where speed and relatively low cost of entry enable rapid start-up and market penetration. Many offer just a branded customer delivery front end to existing products, leveraging market-leading brands, which themselves have taken decades and £Billions to build. Often – as with ordering and delivery of fast food - it represents a mutually advantageous "exploitation" of small-scale, local suppliers who could otherwise not afford to develop the marketing, back-end technology platform, and delivery logistics elements of the product which the new service provides. In doing so, the newcomers often take a greater share of the profit from the 'new product', but little or none of the risk.

Some Enterprises create value through the delivery of niche services based on specialist knowledge, skills or technology, like brand creation, marketing and promotion, or taxation optimisation, which can have particular appeal to smaller Enterprises who don't have the scale to justify such capabilities in-house.

And even when commercial arrangements like these are understood there are can be unappreciated influences at work, which can have seemingly illogical consequences. There is, for example, a growing - and in my mind worrying - trend for services which add no value to the core product or service itself and the consumers of them, but to the investors in those product or service providers. Examples include advisors on tax (avoidance), or legislation and other constraining governance avoidance, and can result in things like relocation of production or perverse routing of transaction processing on grounds other than efficiency.

As the millennium drew towards its end I found myself as Head of IT in a Xerox business Division. We were a modest business unit that had grown out of a capability needed by the parent company. Xerox was an English-speaking company with worldwide markets. Its local maintenance engineers were required to speak English, so service manuals for its products were only needed in English. But User manuals and, more particularly, User Interfaces (operator instruction displays on the machines themselves) needed to be in local market languages, and so needed to be translated. Once we had this translation capability, the company decided to sell the capability as a service to other (non-competing) companies in order to spread the cost. It was a relatively niche market - pidgin English and simple diagrams might be OK for flat pack furniture or low-value products, but not for motor car servicing, pharmaceuticals, military equipment and the like. That was our market - geographically widespread sales of complex, high-value products that required a high degree of accuracy. We translated into about 34 languages, employing native language speakers of each, with relevant subject matter expertise[96] and managed to make about £2m a year from about £20m of external business.

[96] Brexit would make this business untenable now, in view of the need to employ such a wide range of technically skilled linguists in creative ways, to enable sufficient turnover

It was the practice of the company to hold 'kick off' meetings at the start of each year, to share with all staff the results from the previous year, notable events, successes or failures, challenges and plans and targets for the coming year. I was seated on the stage with other members of the Board as the MD explained the expectations of our parent company, and I remember the surprise I felt one year at a question from one of our 200-strong staff: *'but why do we need to make a profit?'* they asked. There was a collective double-take at the time but, with hindsight, it wasn't such a daft question; we were, essentially, producing an essential component of Xerox machines, and any corporate profit ought to be realised at point of sale of those, surely?

The problem was that although Xerox was known to the public as a manufacturer of photocopiers, it was, by then, principally regarded – through its stock market listing - as a technology company[97]. Institutional investors make their investments across a range of business sectors in order to spread their risk, according to their preferred risk profile. Each sector has an expected rate of return, based on how risky and potentially rewarding it is. They would then choose stocks from within their chosen sectors. At that time, technology companies were expected to deliver returns of about 20% - much higher than manufacturers. Although a return of £2m on a turnover of £20m might seem a decent bonus from what would, otherwise, be a corporate cost, at just 10% it was also viewed as a "drag" on the parent's ability to achieve its target 20+%.

---

to keep their mother tongue fresh, and at wages that would be unlikely to meet immigration thresholds

[97] Having been instrumental in the invention of the personal computer, the computer mouse, GUI and WYSIWYG interfaces, the Ethernet, and the laser printer for examples

Investor expectations and simple numbers left us with a problem. It wasn't the fault of anyone in the team, and there didn't appear to be any way of reconciling the 'gap', other than by outsourcing the capability, or at least fully separating the company, by spinning it off from the parent, so that it could be operated within more appropriate investment return expectations. With the benefit of hindsight, maybe that's why Ford, General Motors and our other customers didn't have that essential capability in-house themselves. And presumably it's a problem shared by all companies – the decision to buy-in components or capabilities is based on a more complex reality than just raw price. Any capabilities with a potential return of lower value than that expected from the whole sector are an unwanted "deadweight".

Interestingly, Xerox had huge debts at the time - I was told that they were larger than Hungary's sovereign debt[98] - but it was considered too big and culturally significant in the US to be allowed to fail. I recall, for example, use in the US of the term 'xerox' as a verb, for photocopy, 'team xerox', as a fictional source of plagiarized content and 'the xerox dividend' in relation to personal free use of office photocopying, stationery and other office facilities.

This dynamic helps to explain the explosion of new products and brands developed in western markets, with manufacturing transferred to cheaper, offshore locations. More complex is where elements of an overall 'Value Chain'[99] are outsourced. There can be various devices to

---

[98] Which Wikipedia usefully tells me was 55% of its $146.7bn, or $80.685bn in 2000

[99] 'The value chain describes the full range of activities which are required to bring a product or service from conception, through the different phases of production (involving a combination of physical transformation and the input of various producer services), delivery to final consumers, and final disposal after use' - Kaplinsky, R. and M.

enshrine overall service goals into each contributing part, but you need to recognise that each party may have different overall Enterprise goals, so clear service Goals, Principles, Quality measures and Governance regimes that match your own need to be "baked into" any contracts, sufficient to ensure the integrity of the whole.

Although challenging, understanding a single Enterprise is essential and, even with the "environmental complications" mentioned above, I hope I have shown how it can be achieved in a structured manner.

Now we've nailed the basics, how about using the thought process on a public service Enterprise? Let's look at one that I'm familiar with - policing.

Imagine that we've managed to fully describe policing capabilities, processes, responsibilities, primary strategic objectives, and key measures, leaving just priorities to be set (by the local PCC) on a local basis. Why shouldn't the overall service – 'policing' – for each force be put out to competitive tender? You've got the definition of the service, as the basis for a contract – how is it different to the "contract" that you currently have with your local Chief Constable?

If one force performs particularly well, why shouldn't they be allowed to run the policing of the neighbouring force – say Sussex running, additionally, Surrey, with some cost-saving and operational benefit opportunities? Why not several forces? Why not a foreign police service, working to a clearly defined set of standards and rules, with familiar,

---

Morris (2001), A Handbook for Value Chain Research, prepared for the International Development Research Centre (IDRC)

standardised uniforms worn, and the default language set as English? They could be overseen and regulated by an independent body, such as HM Inspector of Constabulary? How about a commercial supplier? What, if anything, worries you about the idea? What about just a few of the 'support services' which they use – like catering, IT, vehicle maintenance, payroll, Public Relations? How about our Government? Where do you draw the line, and why? I'll talk more later about some public services which you might agree are critical, some of which have been outsourced (privatised), and the consequences of that decision.

A rule of thumb that I find useful is that if something looks like a duck, quacks like a duck and waddles like a duck then it is probably a duck. I also believe that if you keep your eye on the primary purpose then you won't go far wrong. The 'top-down', 'Purpose and Principles' approach can be used to cut through all of the historical baggage and political ideology that often hamper logic by enabling a clear line of sight between intent and action.

And, with that thought in mind, let's take a look at an area that we're all familiar with.

UK PLC

Most people are familiar with the simple principle of public limited companies, where the shareholders employ a board of directors whose task is to grow the value of and return from the company. Things aren't that different at a national level, except that we citizens are the shareholders and we call our board of directors the government. They are entrusted with our national assets and tasked with working to improve the country for our benefit.

You've probably heard the term 'UK PLC' and, thanks to our newly acquired Business Architecture skills, we can consider the country as the big enterprise that it is, applying a similarly disciplined approach to considering change needs and options, and planning how it might best be managed and improved. On the face of it that's a fearsome challenge and, as far as I know, no comprehensive model of the 'Enterprise' in these terms currently exists.

I am not proposing to develop a UK PLC Business Architecture model here. If you were expecting that for the price and size of this book then you have no right to be disappointed. But I hope you can at least entertain the feasibility, and I propose giving a little thought to how such a gargantuan task might be gone about, why the effort might be worthwhile, and what key issues it might help address.

Most of Business Architecture is concerned with facts and logical truths, and can aid understanding of how the whole system functions. UK PLC has, of course, grown organically over centuries, and wasn't built on a formally designed Business Architecture model so, if we want to construct such a model to reflect how things are we need to do it by reverse-engineering. Along the way I expect we'd expose the extent to which politicians have exploited the 'wriggle room' afforded by ambiguity and obfuscation.

Important elements will be the Approach (chiefly its Principles and the Policies & Rules which promote them) and Balanced Scorecard (reflecting the relative value placed on respective elements or perspectives). Not the words used by politicians to describe them, but based on the evidence of hard facts – using comparisons with other, similar states if necessary. We're all too familiar with being told how 'fair', 'compassionate' and 'tolerant' the UK is, for examples, especially when the evidence might suggest otherwise.

Most stable countries, including the UK, ought to be sufficiently longstanding, well-established Enterprises to have a similarly stable set of highest level definitions of enduring Principles and Policies, which reflect the character and culture of the country (Enterprise). Political philosophies, ideologies or dogmas may shift, but usually only tinker at the margins. Beware tinkering with any of these. If, having assessed 'what Principles or values must have been in play to result in the current state of affairs?' leaves you wondering whether the values, culture and character of the country are what you had thought, been led to believe or had hoped, then you should be worried.

Bear in mind that outcomes can depend also on unforeseen circumstances, serendipity, or just plain poor execution, any or all of which can blow a plan off course temporarily, although you'd expect swift recovery or compensatory activity. Be ready also to discover that 'Principles' and Policies might have been articulated purely for marketing purposes (e.g. political manifesto), without there ever having been any intent to actually implement them, been (accidentally or intentionally) vague or ambiguous, or have been well intentioned but poorly planned and/or executed, or overwhelmed by irresistible external factors. All of which provide more 'wriggle room' for politicians and political pundits to exploit. So, when considering competing political manifestos, the relative competence to execute (or perceived competence and trustworthiness), and the global context in which they will need to operate can be as important as the desired outcomes.

So, where to start? The good news is that most of the heavy lifting needed to build a structured model has already been done. Elements are to be found all over the place – government departmental organisation charts, political manifestos, budgets, trade association reports, economic and other forecasts for examples. All that would be needed is collating and linking it all.

Let's assume we want to look at UK PLC as a whole, treating its 67m citizens as shareholders. For the sake of argument, let's say the Head of State – the King – is the Chairman, the PM is Managing Director and the first duty of the board of directors (the government) is to ensure the protection of the country and the wellbeing of its (shareholder) citizens.

As in a corporate, commercial environment, management isn't about doing everything, but about making sure everything is done. At the same time <u>responsibility</u> cannot be outsourced or delegated. The UK is, of course, a very 'mixed economy', with the provision of goods and services coming from both public and private sectors. This approach recognises that the Government don't need to provide goods and services necessary for the functioning of the country, but they do retain responsibility for ensuring that those goods and services are provided for the benefit of its (shareholder) citizens. Fortunately, there is a plethora of experience and mature existing sources in the UK on which to draw.

Bear in mind that a commercial provider to whom services are 'outsourced' cannot have the same primary driver as an 'in-house' provider. They are legally obliged to have as their primary objective the best outcomes (company growth, retained and distributed profits) for their own shareholders - not the citizen shareholders for whom they will be doing the work. Commercial parties that take over services might not even have any corporate allegiance to the country, and may employ various devices to avoid paying local taxes, for example. Regardless, the fundamental difference in motivations between public and private providers is rarely properly recognised within privatisation debate.

The 'public sector' is still probably the biggest 'organisation', and with some fairly clear, consistent and accessible documentation describing individual services and the integration between them, so that would seem to be a good place to start. Although it's not a single, homogenous 'Enterprise', its central constituent parts ought to at least share common Principles, and undertaking this exercise might highlight opportunities to develop and use standard templates which promote the benefits of systematically recording the important links between, say, Processes, Rules, Values, Principles and Policies – as you would expect in any sizeable, enlightened commercial Enterprise.

A quick search of the Web reveals the current organisation and scale of UK central government[100]

[100] https://www.gov.uk/government/organisations

There are 23 Ministerial departments

Attorney General's Office
Cabinet Office
Department for Business, Energy & Industrial Strategy
Department for Digital, Culture, Media & Sport
Department for Education
Department for Environment Food & Rural Affairs
Department for International Trade
Department for Levelling Up, Housing & Communities
Department for Transport
Department for Work & Pensions
Department of Health & Social Care
Foreign, Commonwealth & Development Office
HM Treasury
Home Office
Ministry of Defence
Ministry of Justice
Northern Ireland Office
Office of the Advocate General for Scotland
Office of the Leader of the House of Commons
Office of the Leader of the House of Lords
Office of the Secretary of State for Scotland
Office of the Secretary of State for Wales

20 Non-ministerial departments

The Charity Commission
Competition and Markets Authority
Crown Prosecution Service
Food Standards Agency
Forestry Commission
Government Actuary's Department
Government Legal Department
HM Land Registry
HM Revenue & Customs
NS&I
The National Archives
National Crime Agency
Office of Rail and Road
Ofgem
Ofqual
Ofsted
Serious Fraud Office
Supreme Court of the United Kingdom
UK Statistics Authority
The Water Services Regulation Authority

412 Agencies and other public bodies

From Academy for Social Justice
via
Companies House
to
Youth Justice Board for England and Wales

108 High profile groups

From Bona Vacantia
via
Fleet Air Arm Museum
to
Youth Custody Service

13 Public corporations

Architects Registration Board
BBC
BBC World Service
Channel 4
Civil Aviation Authority
Crossrail International
Historic Royal Palaces
London and Continental Railways Limited
National Employment Savings Trust (NEST) Corporation
Office for Nuclear Regulation
The Oil and Pipelines Agency
Ordnance Survey
Pension Protection Fund

And 3 Devolved administrations

Northern Ireland Executive
The Scottish Government
Welsh Government

Even then, regional, locally-managed and -controlled operations often have a local democratic mandate, which might be at odds with national interests. Especially when local democratic bodies have principal responsibility for service provision, but are heavily reliant on central funding, that might be subject to conflicting political influence. In that case there is no common purpose, and potential for conflict. Either central government needs to be fully accountable for all local services (education, health, care, policing, potholes and so forth), or local bodies need to receive the funding for those services direct. In the UK we have a mixed system, which enables central government to 'save money' whilst allowing the resultant pain to be perceived as being the fault of local responsible providers.

I think you'd need to go through a number of iterations and layers of information to be gleaned from the broad array of existing sources, just to establish the basic structure of a model. It would seem sensible to establish early on whether the intention was to provide an information platform onto which existing sources could transfer, or the necessary standards to make future updating less onerous: data scope, format and ownership, access permissions and control, updating responsibility, back-up regimes and so forth.

Next could be a list of locally delivered public services and products, which could be gathered from a representative selection of Local Authorities' web sites and publications.

Charities, Social Enterprises and other 'not for profit' organisations often plug gaps which have been left. There were approximately 169,000 charities registered by the Charity Commission[101] in England and Wales in 2021.

Foodbanks are one example. The number of foodbanks in the UK had grown to over 2,200 by July 2021[102], for example, and about 2.5m food parcels were delivered

[101] https://www.gov.uk/government/organisations/charity-commission/about
[102] https://commonslibrary.parliament.uk/research-briefings/cbp-8585/

through the Trussell Trust's 1,300 Foodbanks in the year 2020/21[103].

Then, of course, you need to add the range of products and services that are provided by commercial companies, and for which a variety of sources are available. There is a government Standard industrial classification of economic activities (SIC)[104], but you could also look to others such as the British Chambers of Commerce[105], or even Amazon or eBay listing categories. When all available conventional sources had been tapped, creative thought is likely to be needed, and consideration of other perspectives could usefully be taken to validate the picture. How about considering an average citizen's daily life experience? What better memory-jogger than the Beatles' 'A Day in the Life'[106]

♬ ***I read the news today, oh boy*** ♬
The press, Internet newsfeeds, TV and Radio, the media generally

♬ ***About a lucky man who made the grade***
And though the news was rather sad
Well, I just had to laugh

I saw the photograph ♬
Image distribution

♬ ***He blew his mind out in a car*** ♬
Car production, maintenance, fuel, licensing, insurance

♬ ***He didn't notice that the lights had changed*** ♬
Traffic management, road maintenance

---

103 https://www.trusselltrust.org/

104 https://www.gov.uk/government/publications/standard-industrial-classification-of-economic-activities-sic

105 https://www.britishchambers.org.uk/

106 John Lennon / Paul McCartney

♫ ***A crowd of people stood and stared***

They'd seen his face before

Nobody was really sure if he was from the House of Lords

I saw a film today, oh boy ♫

Entertainment industry (including venues and other distribution channels)

♫ ***The English Army had just won the war*** ♫

The whole defence industry (weapons design, manufacture, procurement, testing, storage and deployment, surveillance and intelligence)

♫ ***A crowd of people turned away***

But I just had to look

Having read the book ♫

Publishing

♫ ***I'd love to turn you on***

Woke up, fell out of bed ♫

Furniture and linen, laundry equipment, products and services

♫ ***Dragged a comb across my head*** ♫

Personal grooming

♫ ***Found my way downstairs and drank a cup*** ♫

Accommodation, food and beverages, small household appliances, crockery

♫ ***And looking up, I noticed I was late*** ♫

Employment

♫ ***Found my coat and grabbed my hat*** ♫

Clothing

♫ ***Made the bus in seconds flat*** ♫

Public transport

♬ ***Found my way upstairs and had a smoke*** ♬
Tobacco/Vapes

♬ ***And somebody spoke and I went into a dream***
Recreational drugs (?)

♬ ***I read the news today, oh boy***

Four thousand holes in Blackburn, Lancashire ♬
Policing[107]

♬ ***And though the holes were rather small***

They had to count them all

Now they know how many holes it takes to fill the Albert Hall ♬
Statistical and other analysis

♬ ***I'd love to turn you on*** ♬

Whatever route you take, aiming to cover all the needs of citizen shareholders, and however deep you go, your Capability Model for UK PLC will almost certainly miss key elements. As an example, the recently discovered critical shortage of CO2 in the UK which, it turned out, is vital not only in making fizzy drinks and pumping them to customers in bars, but also in the animal slaughter process, promoting plant growth, food packaging and preservation, and probably much else. Almost too late it was realised that 60% of the country's supply came from a single (foreign-owned) fertiliser manufacturer[108] that had decided to up sticks and close its UK manufacturing, since rising energy costs had rendered their operations

[107] I believe the reference was to the findings of a study into the number of drug addicts in Blackburn

[108] CF Industries

commercially unviable. So the UK government stepped in with taxpayer funds to subsidise their operations for a period, hoping that the market would adjust.

Once you are confident that you have a sufficiently comprehensive Capability Model, you can use it to clarify who fulfils identified needs - Central government, local authorities, private commercial enterprises, charities, self-help and so on. As described earlier, it could also help to identify 'hot spots', strengths and weaknesses (or gaps and opportunities).

Now you can see the importance of a clear understanding of the capabilities required to supply every service (or product) needed. It's worth showing again the relevant parts of the diagram which I introduced earlier under general Business Architecture:

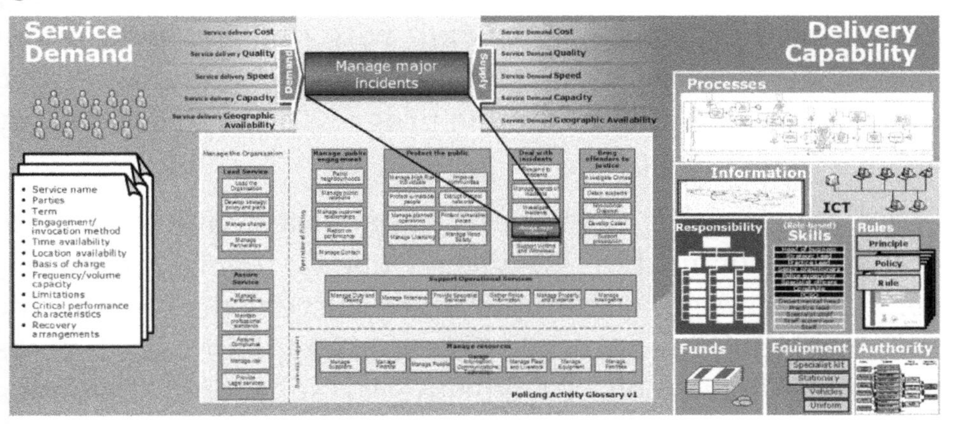

Specifically, you need to ensure that you have, for each and every required national capability, appropriate and adequate

- Information (access, the right to use, analytic capabilities);
- ICT (communications infrastructure, IT systems and equipment);

- Assigned Responsibilities (and accountabilities);
- Skills (sufficient well-trained and competent people);
- Rules (including legislation, standards, and authoritative oversight);
- Funds (budgets, cash flow, authority);
- Equipment (including vehicles, buildings, specialist equipment);
- Authority (any necessary Intellectual Property and other rights and powers to act).

The model could also be used, together with financial and timescale forecasts, as the foundation for strategic government planning and budgeting, political manifestoes, and to make clear the weaknesses and opportunities that specific changes are intended to address.

All services ought to be focused on achieving the primary Vision for the country, consistent with its Principles, and it should be possible to trace them back to that Vision and those Principles, and to enable progress resulting from any change to be measured.

Interdependencies between capabilities necessary to realise recognisable services need to be carefully considered. A good example is the current problem in the UK of Ambulance waiting times. Complaints and press headlines have led to knee-jerk 'solutions' involving Ambulance service funding or structuring, but the true cause is more often that Ambulances need to queue for hours at hospital A&E receptions due to a lack of available hospital beds or assessment staff. Patients are safer in the care of medical professionals but, whilst Ambulances wait, they are not available for fresh calls. Many hospital beds are occupied by patients who aren't in need of hospital treatment or care, but there is no available capacity at a care home, or support for them in their home. The problem manifests itself 'downstream' in the form of long waits for an Ambulance, whilst the solution may lie in improvement of social care services.

Even if you don't plan to outsource (or privatise) a capability or service, considering and documenting it as though you did is a useful discipline. 'Supply contracts' can be between government departments and can form the basis of "hand-over" agreements between parties involved in shared 'enterprises', and of performance incentives. You need to set appropriate (Balanced Scorecard-based) quality standards in relation to the overall, meaningful products or services and to include sufficiently compelling consequences to ensure that's what is delivered, together with a strong, timely audit and governance regime to assure actions. If, for whatever reason, you ever need to end the arrangement and bring an outsourced service back "in house", think about how easy would that be to do, and what issues would require resolution. And bear in mind that the more parties involved in a service, the more hand-offs there will be, the greater potential for errors and increased dilution of accountability, or opportunity for those with responsibility to try to deflect.

Going through this thought process – about the whole service, or a large component of/capability within a business – can be a useful aid to thinking about any Enterprise. It IS Business Architecture.

In a complex mixed economy, such as the UK, the executive still has vital role to play in steering the Enterprise for the benefit of the citizen shareholders who employ them. But, when I hear proponents of 'small government' and 'free, open markets' I am reminded of my Father watching football on the television. When a player was deemed to not be offside 'because he was not interfering with play' his retort would always be 'then why is he on the pitch?' Fair point, I always thought.

Wriggle room

I mentioned 'wriggle room' earlier. Some is accidental and some deliberately engineered. Often it results from an appeal to public imagination whilst using ambiguous or misleading language. There were clear examples in the VoteLeave Brexit campaign manifesto such as 'Take Back control of our Borders', when we already had control, except on our only UK-EU land border, in Ireland, and for which no practical plan to achieve it has been proposed in almost seven years following the referendum vote. Words are important. Particularly with consequential matters they should be used with precision, with measurable intentions, all of which need to be understood by all stakeholders.

Let's take energy as an example.

UK energy

The UK took a leading stance on the switch from fossil fuels to renewable energy when it hosted, as President, the 26th UN Climate Change 'Conference of the Parties' (COP26) in Glasgow between 31 October and 13 November 2021 and, in February 2022, when Russia invaded Ukraine, joined the increasing global rhetoric about the virtues of national energy independence. So it was no surprise that these issues featured highly in the 'British Energy Security Strategy (BESS) – Secure, clean and affordable British energy for the long term' published[109] in April 2022. Even when we know the ideal destination, we might not be in the best place to start so it's vital to begin with clear analysis and strategy.

In his Foreword to that document the Prime Minister noted that the UK had 'drifted into dependence on foreign sources' and went on to say that 'if we're going to get prices down and keep them there for the long term, we need a flow of energy that is affordable, clean and above all, secure. We need a power supply that's made in Britain, for Britain – and that's what this plan is all about.' And whilst the PM recognised that 'we can't simply pull the plug on all fossil fuels overnight' he also signalled the increasingly important role that hydrogen would play: 'We're going to produce vastly more hydrogen, which is easy to store, ready to go whenever we need it, and is a low carbon superfuel of the future'. This final point recognises the fact

---

109 https://assets.publishing.service.gov.uk/government/uploads/system/uploads/attachment_data/file/1069969/british-energy-security-strategy-web-accessible.pdf

that energy consumption tends to be geographically remote from and asynchronous to generation, so storage and transmission media also needs to be considered.

On the face of it that all sounds pretty good, and suitable to serve as a 'Balanced Scorecard' for UK energy against which to measure specific initiatives and progress. Until you read it forensically, notice a few key omissions, and realise that what we have here is an example of 'wriggle room' being "baked-in", ready to exploit when assumed (but probably never intended) targets are not met.

- Made in Britain, for Britain

  (For me this is a big red flag. I would have expected 'Made **by** Britain, to meet all of Britain's needs and, if possible, provide a new national income stream'. I am sure that this wasn't accidental, as we'll see later)

- Secure, long-term supply

  (I take this to reflect the experience of the petroleum crises of the 1970s and more recently the effects of dependence on Russian gas following their invasion of and war with Ukraine, and to mean secure in terms of not being reliant on other national powers or commercial suppliers. Let's also assume that it implies a reliable supply adequate to meet current and future national needs)

- Affordable

  (Presumably with prices under national control, or at least not under the control of other nations or commercial parties, nor subject to the vagaries of global markets)

- 'Clean'

  (An interestingly vague term, especially since the body of the report refers to 'renewables' which would include, for example, the bioenergy burning of purpose-grown timber, as in the Swedish model[110]. Timber is central to Sweden's economy – Bioenergy - most of it from forests - accounts for three-eighths of all the energy used in Sweden and the forestry industry employs over 60,000 people directly and is indirectly responsible for around 200,000 jobs. In 1993, Sweden made its forests "a national resource". Each year around 120 million cubic metres of forest grows and around 90 million cubic metres of that growth is harvested, so national stock is growing (pun intended). It has doubled in the last century. And more carbon is captured than released. Does that count as 'clean'? And is nuclear energy 'renewable'?).

One element of the 'Ten point plan' contained within the BESS strategy is 'Driving the growth of low carbon hydrogen' but this turns out to be a fairly tame model for public underwriting of supply-side (hydrogen production) risk. That view is reinforced by another of the 'Ten points' - 'Accelerating the shift to Zero Emission Vehicles' which speaks only of historic investments in electric vehicles and supporting infrastructure. Meanwhile, the only mention of hydrogen-powered vehicles relates to buses.

So does government action support those claimed 'strategies'?

---

[110] https://www.irena.org/-/media/Files/IRENA/Agency/Publication/2019/Mar/IRENA_Swedish_forest_bioenergy_2019.pdf?rev=5b9d9f8da56c438b97be3eeedba40e5a

Obvious first steps need to be to minimise consumption, as far as possible. Walk more. Cycle rather than drive. Use public transport more. Put on a woolly when you get chilly. Practical steps could include insulation of homes through revised building regulations for new homes and grants for retrofitting to existing ones, reduced heating through lowered thermostats, increased efficiency of lighting and appliances in domestic, commercial and public facilities, and more efficient, subsidised public transport. But whilst, in the BESS strategy, the UK government recognises these issues, responsibility is left to the general public: 'this is not being imposed on people and is a gradual transition following the grain of behaviour. The British people are no-nonsense pragmatists who can make decisions based on the information'. It's true that market forces will tend to operate, as costs drive down demand, but change can be given impetus with some positive government intervention (the BESS mentions some grants for household heat pumps, for example). Excluding changes to Building Regulations seems to have been a strangely missed opportunity.

In contrast, France has, for example, introduced a number of practical initiatives in recent years (albeit that some were prompted, initially at least, by a need to improve road safety, or as a result of problems with their nuclear power plants). In 2018, the speed limit on single-carriageway A and B roads in France was reduced from 90km/h to 80km/h, and those limits are supported by a proliferation of very advanced speed enforcement cameras[111] (which collected €859million in fines in 2021). They have recently

[111]' France's new traffic camera is so powerful, it can spot if you're not wearing a seatbelt' https://www.thetimes.co.uk/article/drivers-beware-le-super-radar-knb2vf702?gclid=Cj0KCQiA-JacBhC0ARIsAIxybyNZPcKB2_xzy2VdSjy9K-vGBuSKAFLYUyp-K6UOTuAoxh6XouxjQrQaAsHKEALw_wcB

approved legislation that will require all car parks with more than 80 spaces to install solar panel roofs, as part of a wider programme that will see solar panels occupy derelict plots, vacant land alongside roads and railways, and some farmland. This is expected to add 11 gigawatts to the French electricity grid equal to 10 nuclear reactors. That may be ambitious[112], but such solar panel-roofs are becoming common on agricultural barns and in supermarket car parks. The switch to domestic electric smart meters ('Linky') is almost complete, incentivised by charging different electricity rates, signs announcing 'Extinction eclairage public de 0h00 a 5h00' have sprung up in towns nationwide and the government have also recently introduced a range of state aid for heat pump installations.

All initiatives, however small, will contribute to significant overall reductions, benefitting individual households, businesses, and the national economy.

When it comes to energy generation, there isn't a one-size-fits-all model for all countries. Some have a wealth of fossil fuel resources – coal, oil, and gas - or uranium. Some have plentiful geothermal energy, or an abundance of sunlight or wind. Others have coastlines with strong tidal flows, or swathes of renewable forest. And many, of course, have a history which includes colonial possessions where such resources could be easily had, or the military might which enabled them to be commandeered at will.

---

[112] https://www.cittimagazine.co.uk/comment/frances-plan-for-solar-panels-on-all-
The shift in fuel mix has been fairly dramatic, over the last three decades, with nuclear power being relatively stagnant, and renewables only recently increasing to fill the gap left by coal.
car-parks-is-just-the-start-of-an-urban-renewable-revolution.html

Other issues to consider include pollution, health and safety associated with extraction and processing, waste disposal, decommissioning costs, the desirability to have a mix of types, and incidental costs/benefits, such as social amenity creation and skill, job and market development, especially where there are global opportunities in innovative growth industries.

The UK has as chequered history as any. Our pollution-stained building facades and slag heaps are testament to the decades of dependence on coal. After a steady decline in demand the last operating deep coal mine closed in December 2015. Coal imports rose for 20 years from 1995[113] in order to maintain supplies for electricity generation but the contribution of coal to the national electricity grid has been in sharp decline since 2015 and, in November of that year, the UK Government announced that all the remaining fourteen coal-fired power stations would be closed by 2025. Coal was not even deemed worthy of inclusion in the BESS strategy and, as I write, the Secretary of State for Levelling Up, Housing and Communities and Minister for Intergovernmental Relations' (?) has been at pains to explain that he has just approved a new coal mine not for energy, but in order to supply the steel making industry, as an alternative to current supplies from Russia. Curiously, we have already ceased importing coking coal (the kind used for steel production) from Russia, making up the shortfall by increasing imports from the EU, Australia, Canada and the USA[114], so it isn't a simple binary choice

---

[113] UK ENERGY IN BRIEF, 2021 - Department for Business, Energy and Industrial Strategy (BEIS) - https://assets.publishing.service.gov.uk/government/uploads/system/uploads/attachment_data/file/1032260/UK_Energy_in_Brief_2021.pdf

[114] Coal Imports, BEIS.Gov.UK, 29 September 2022

between Russian and new British supply. Nor does the claim that the decision is environment-based – that it's better to mine our own than to 'export the problem overseas', hold water, since 85% of the new coal will, apparently, be exported.

In 2019 the UK produced 7 million tonnes of steel, whilst China produced 996 million tonnes[115].

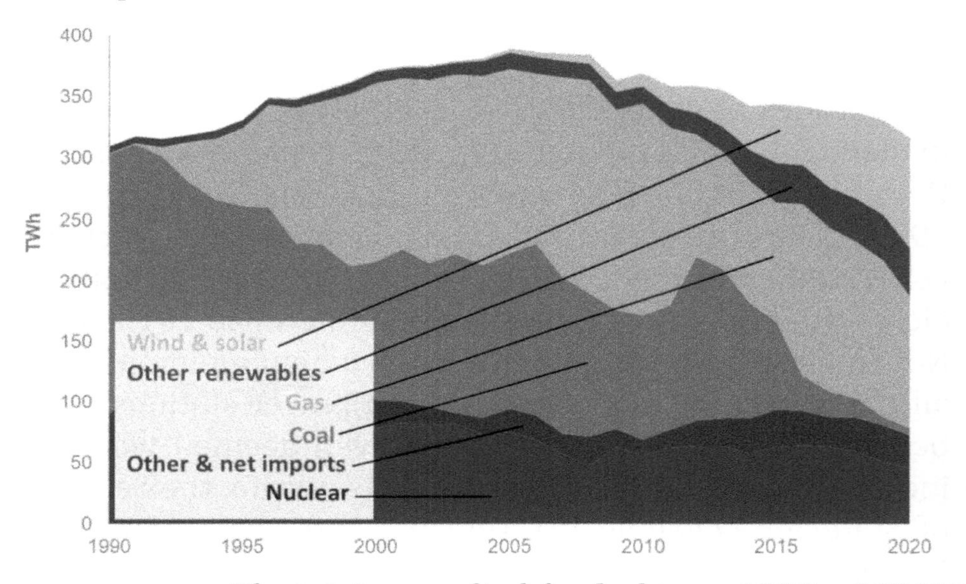

Electricity supplied, by fuel type, 1990 – 2020[116]

For interest, steel is increasingly reused and British Steel is now owned by a Chinese firm.

We enjoyed a couple of decades between 1980 and 2005 as net exporters of oil and gas, but are now heavy importers once again.

---

[115] https://commonslibrary.parliament.uk/research-briefings/cbp-7317/

[116] From 'UK energy in brief, 2021' - (BEIS)

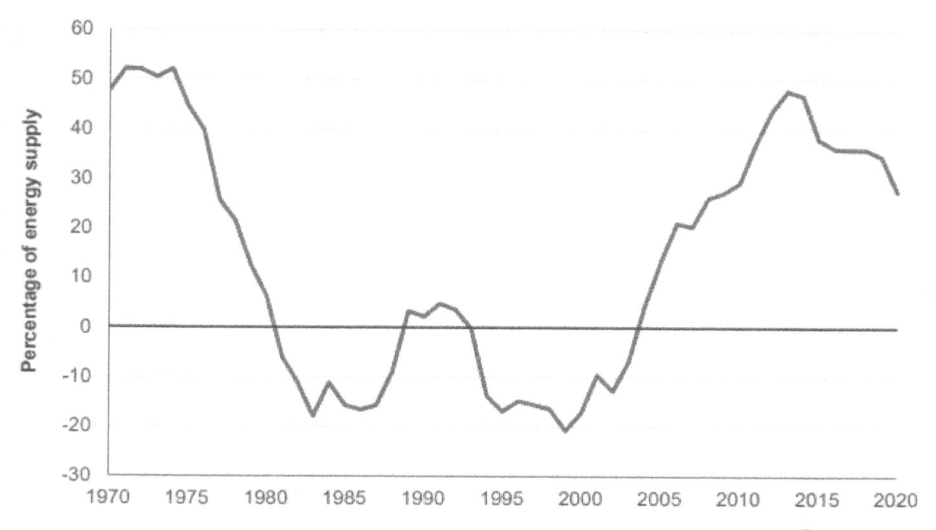

Percentage

	2000	2005	2010	2018	2019	2020
Coal	39%	71%	52%	78%	68%	47%
Gas	-11%	7%	40%	50%	50%	47%
Oil	-55%	-3%	14%	29%	26%	10%
Total	**-17%**	**13%**	**29%**	**36%**	**35%**	**28%**

Import dependency, 1970 – 2020 (From 'UK energy in brief, 2021' - (BEIS))

Most (64%) of the gas consumed in the UK is used for domestic heating[117]. Half is imported – mostly (77%) from Norway[118].

Interestingly, Norway took a very different approach to that of the UK following discovery of oil and gas in the North Sea. One of the overall principles of Norway's management of its petroleum resources is that 'exploration, development and production must result in maximum

[117] 'Subnational Electricity and Gas Consumption Statistics, Regional and Local Authority, Great Britain, 2020, Department for Business, Energy and Industrial Strategy (BEIS) 23 December 2021

[118] https://www.ons.gov.uk/economy/nationalaccounts/balanceofpayments/articles/trendsinukimportsandexportsoffuels/2022-06-29

value creation for society, and that revenues must accrue to the Norwegian state and thus benefit society as a whole', so it created a sovereign wealth fund (State's Direct Financial Interest (SDFI)) – now the largest sovereign wealth fund in the world - and puts its (net) oil revenues into the Government Pension Fund. The effect is profound, and enviable. Norway PLC seems to take a very different approach to its national assets and its (citizen) shareholders, and to have different principles in relation to its national interest.

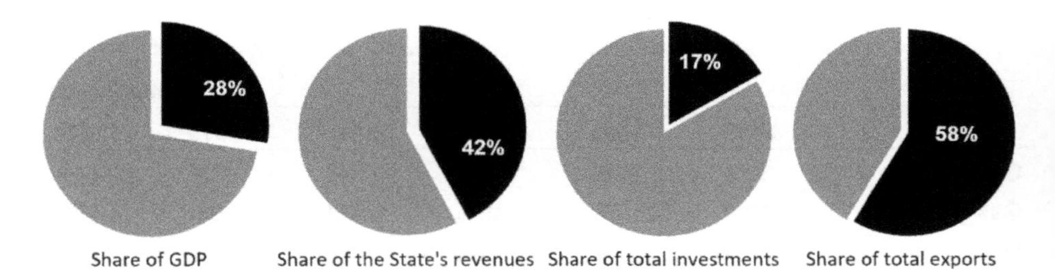

Macroeconomic indicators for the Norwegian petroleum sector, 2022[119]

Nuclear power is increasingly seen as an essential component of a non-fossil fuel energy strategy. The world's first full scale nuclear power station opened on October 17, 1956, at Calder Hall, in Cumbia. France didn't open its first until 1962, and yet now has over 50, generating 69% of its electricity. After a relatively late start, lagging third behind the USA and France in installed nuclear capacity[120], and a 2-year hiatus as a result of safety concerns following the Fukushima meltdown in Japan, China recently announced plans for a further 150 reactors - more than the rest of the

[119] https://www.norskpetroleum.no/en/economy/governments-revenues/

[120] Energy Monitor, 20 December 2021 https://www.energymonitor.ai/sectors/power/weekly-data-chinas-nuclear-pipeline-as-big-as-the-rest-of-the-worlds-combined/

world has built over the past 35 years – at a cost of $440bn. The cost of nuclear-generated power has always been regarded as relatively inexpensive. In 1954, Lewis Strauss, who chaired the US Atomic Energy Commission, famously predicted that nuclear energy would make electricity "too cheap to meter". Whether or not that prediction was borne of a nuclear arms race remit to encourage the nuclear industry and its by-products, it seems, with hindsight, to have been spectacularly optimistic.

Uranium is relatively plentiful, with Kazakhstan being the major producer (39% of world total) followed by Canada (22%) and Australia (10%). It is a commodity traded on global markets and is a relatively cheap fuel, compared to fossil fuels. Only 5% of that used by the USA is from the USA (interestingly 14% comes from Russia). Ours comes via an American company (Westinghouse). The expected increase in demand resulting from China's recently announced construction programme will probably result in a rise in the price of the commodity, but is likely to be easily met by increased production.

Here in the UK, 66 years since we began, modular nuclear reactors, which are faster to build and carry fewer security risks, have been proposed, but the technology is still nascent. Rolls Royce has been touted as a leader in the field but, although the UK government have made encouraging noises, it is Qatari investment that is backing it. Of the conventional nuclear power fleet, only about 16% of our electricity is generated in 9 nuclear power stations, which are all operated by EDFE (part of the French state electricity company, EDF). That percentage is declining, as the seven Advanced Gas-cooled Reactors (AGR) are nearing the end of their operating lives, and are already between three and seven years behind planned decommissioning dates.

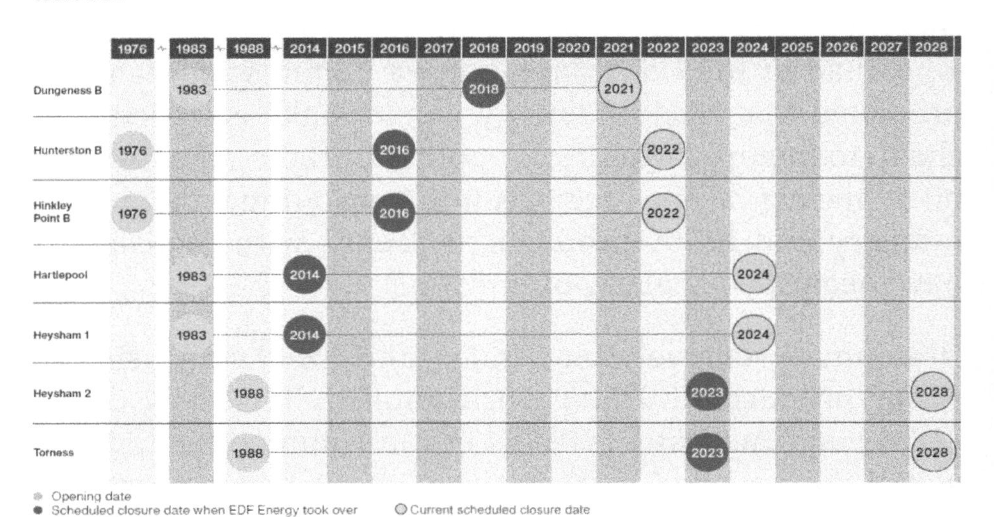

Original and revised closure dates of the AGR stations [121]

The UK government got the headlines they were looking for from the April 2022 'British Energy Security Strategy (BESS)' regarding nuclear power, with the PM's 'We're bringing nuclear home, with one nuclear plant, one nuclear reactor, every year for eight years, rather than one a decade.' Most of the press[122] chose to interpret (and report) that as 'eight nuclear stations in eight years', suggesting that the gap being left by decommissioning of existing sites would be plugged. Less attention was paid to the fact that the PM was talking only about **approval** of plans, and nuclear industry bosses' acknowledgement that

[121] Source: National Audit Office analysis of information provided by EDFEnergy and the Department for Business, Energy and Industrial Strategy – National Audit Office: The decommissioning of the AGR nuclear power stations, 28 JANUARY 2022 - HC 1017 https://www.nao.org.uk/wp-content/uploads/2022/01/The-decommissioning-of-the-AGR-nuclear-power-stations.pdf

[122] An example, from the Mirror, 7/4/2022 https://www.mirror.co.uk/news/politics/boris-johnson-pledges-up-8-26658631?utm_source=linkCopy&utm_medium=social&utm_campaign=sharebar

the new nuclear power stations will take well over a decade to build. Nor was reference made to the history. Such as the fact that it was sixteen years since the independent consultants' report[123], commissioned by Ministers, offered a list of the most suitable sites, fourteen years since Business Secretary John Hutton told MPs that 'A new generation of nuclear power station would give a "safe and affordable" way of securing the UK's future energy supplies while fighting climate change'[124], and since the then PM, Gordon Brown, called for eight new nuclear plants to be as part of a 'nuclear renaissance' in the UK. Later that year the French state-owned energy giant EDF negotiated a £12.4bn deal to buy British Energy, which ran eight nuclear sites, together with land on which new reactors could be built. The deal was completed the following year (2009) for £12.5bn. The UK government gave the green light[125] to eight new reactors on 18 Oct 2010, with Chris Huhne MP, Secretary of State for Energy saying 'I'm fed up with the stand-off between advocates of renewables and of nuclear which means we have neither. We urgently need investment in new and diverse energy sources to power the UK'. It's incredible to see a further announcement, twelve years later, suggesting that it represents a new initiative - perhaps even 'urgent' in response to recent Ukraine war-induced global energy prices.

EDF negotiated a guaranteed fixed price – a "strike price" – that it will be paid for electricity from Hinkley Point C under a government "Contract for difference" (CfD). The

[123] Jackson Consulting, April 2006
[124] Reported on BBC News, 10/1/2008
http://news.bbc.co.uk/1/hi/uk_politics/7179579.stm
[125] https://www.gov.uk/government/news/huhne-highlights-urgent-need-for-new-energy

price is £92.50/MWh (in 2012 prices), which will be inflation-adjusted (£106/MWh in 2021) during the construction period and over the subsequent 35 years tariff period. The plant was expected to be completed in 2023 and to remain operational for 60 years.

In 2015, China General Nuclear (CGN) signed a Strategic Investment Agreement to participate in three nuclear projects in the UK: Hinkley Point C, Sizewell C and Bradwell B. CGN has funded 33% of the project so far, with EDF holding a 67% stake. Completion is expected in 2026, and already costs are running 44% higher than expected in 2015[126].

As of May 2022 Hinkley Point C was two years late and the expected cost had increased to £25–26 billion - 50% more than the original budget from 2016[127]. But although the initial cost of building nuclear power stations is high, it is not the biggest cost. The claimed lifespan of the new Hinkley Point C station is 60 years but, according to the International Atomic Energy Agency (IAEA) the operational life-span of a nuclear power station is only 20 to 40 years[128]. And decommissioning costs – which fall, ultimately, to UK taxpayers - are eye-wateringly high. We have the experience of our first generation of stations as a guide. When, in 2009, the UK government sold its stake in British Energy to EDF Energy (EDFE) the decommissioning agreements were not a major focus (simply acknowledging that EDFE would act as, effectively a "cost plus" contractor, with the UK government retaining responsibility for underwriting EDFE's costs). New agreements were signed

[126] China Research Group https://chinaresearchgroup.org/research/briefing-chinas-involvement-in-uk-nuclear-power
[127] Wikipedia https://en.wikipedia.org/wiki/Hinkley_Point_C_nuclear_power_station
[128] https://www.iaea.org/sites/default/files/29402043133.pdf

in June 2021, aiming to reduce the UK's liability by clarifying what constitutes qualifying decommissioning costs, and introducing reward incentives for EDFE to accelerate the removal of fuel and make other cost savings, reflecting the fact that post-operational nuclear power stations need managing, and that it is more expensive to manage a fuelled one than one from which the fuel has been removed. EDFE has estimated that the fixed costs to manage and maintain a station that is not generating electricity but still holds fuel could amount to around £140 million per station per year, reflecting the cost of having to maintain the station, with all the associated staffing and facilities required to keep the reactor in a safe state. These fixed costs are expected to fall significantly to around £25 million - £35 million per station per year once the fuel has been removed. This amounts to a tidy sum when talking about 7 stations and many years to decommission.

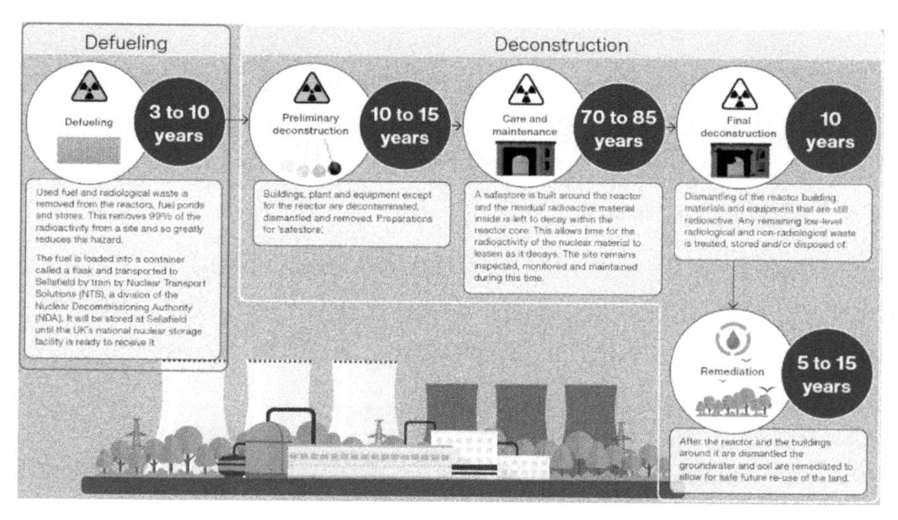

EDFEnergy's indicative process and timeframe for decommissioning the AGC reactors[129]

[129] Source: National Audit Office analysis of information provided by the Nuclear Decommissioning Authority and EDFEnergy - National Audit Office: The

The most recent report[130] by the National Audit Office recognised the EDFE-calculated cost of decommissioning the seven UK AGRs at £23.5bn. That figure doesn't take account of waste disposal (long-term nuclear waste storage/disposal remains a major conundrum) and other costs, so is generally regarded as wide of the mark for total decommissioning costs. In fact the Guardian reported[131] in September 2022 that the cost could be as high as £260bn, and the exercise will take 100 years. That figure is based on the Nuclear Decommissioning Authority (NDA)'s own estimate of £149bn. Add to that the original construction, commissioning, operation and fuelling costs for an energy supply peaking at less than 20% of national demand and it doesn't sound 'too cheap to meter' now.

Perhaps the UK fares better in the other 'low carbon' sectors, where there are significant commercial opportunities for market leaders?

In 2020 renewables accounted for more than 43.1% of the UK's total electricity generated, outstripping fossil fuels for the first time in the nation's history.[132]. Let's look briefly at how that is made up.

Sunshine in the UK might best be described as capricious. Solar power now contributes just 1.8% to our energy grid, but at least that figure is growing – 24% up in Q4 2021 on

---

decommissioning of the AGR nuclear power stations, 28 JANUARY 2022 - HC 1017 https://www.nao.org.uk/wp-content/uploads/2022/01/The-decommissioning-of-the-AGR-nuclear-power-stations.pdf

[130] HC 1017, 28/01/2022 https://www.nao.org.uk/reports/the-decommissioning-of-the-agr-nuclear-power-stations/

[131] https://www.theguardian.com/environment/2022/sep/23/uk-nuclear-waste-cleanup-decommissioning-power-stations

[132] https://www.nationalgrid.com/stories/energy-explained/how-much-uks-energy-renewable

the previous year. Ten of the world's top 14 solar panel manufacturers are Chinese, two are from the USA, one from Canada and one from South Korea. In Europe, German manufacturers dominate. It's not a key UK competence.

The UK is generally accepted as being a relatively windy country, so it's not surprising that so much of our electricity is now generated by wind turbines, nor that by 2019 the Economist was describing Britain as 'the world's biggest offshore wind market'. Whilst offshore wind generates just 2% of global renewable power, it now accounts for 27% of the UK's total power. Although the UK seems to be benefitting from plenty of "greener" energy, we don't seem to have carved ourselves a lead role in the market; most of the kit seems to be made by Siemens, and Ørsted – a majority Danish state-owned company, have emerged as world leaders - the largest offshore wind farm company in the world.

But wind is fickle. As the (BEIS)' 'UK energy in brief, 2021' noted

> ***Electricity generated from renewable sources increased by 13% between 2019 and 2020. The large increase is mostly due to favourable weather conditions, as installed capacity grew only marginally. Total wind generation increased by 18% thanks to exceptionally strong wind speeds. Wind generation was particularly high during Quarter 1 of 2020, when storms Clara and Dennis hit the UK. Average onshore wind speeds in 2020, at 9.1 knots, were 0.8 knots higher than in 2019***

'Bioenergy' – contributes more through fuelling electricity generation, direct heat generation, and road transport than onshore wind turbines, but doesn't sound as sexy, so gets less press coverage.

Of all the natural energy that the UK is blessed with, surely the tides must be the most valuable? It's easy for anyone born here to take them for granted, but our common tidal ranges are exceptional. The one local to me is over 6m - twice a day, regular and reliable, regardless of sunshine or wind and, thanks to the moon, predictable for centuries to come.

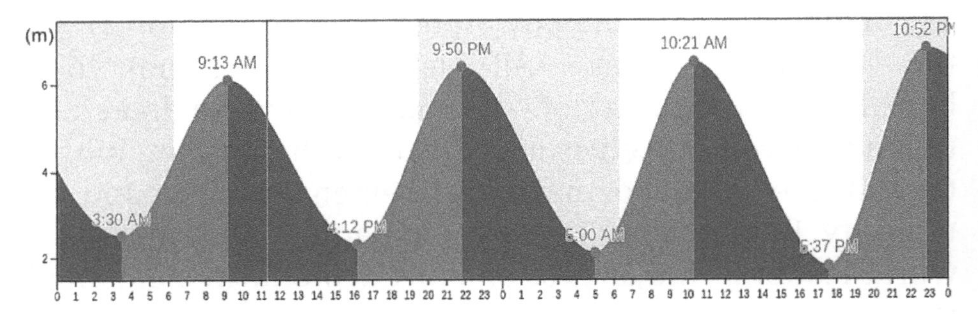

Chart of my local tides[133]

It's 4-5m all around most of the UK, and peaks at over 14m in Bristol. I recall the faces of Italian visitors when they saw, and struggled to comprehend, the difference between low- and high-tide lines. Not surprising, when you consider that the tidal range in, say, Palermo, is about 0.3m. And it's not just the Mediterranean. In Rio de Janeiro, Brazil, Lima, Peru, Wellington, New Zealand, and Iwaki, Japan it's about 1m. In Fremantle, Western Australia it's even less. In Cape Town, South Africa it's about 1-2m and Mumbai, India about 2-3m.

In January 2021 Ovo energy described Tidal power as 'a giant on the horizon of renewable energy'[134]. And nowhere

[133] From https://www.tideschart.com/

in Britain is further than 120Km from the sea, meaning that generated power never needs to be carried far.

A private company - Tidal Lagoon Power (TLP) - submitted plans for a trial tidal lagoon in Swansea Bay in February 2014[135]. It promised to be the world's largest power-generating lagoon, to begin generating power by 2022, and the first of six proposed.

Artist's impression showing how the lagoon would be designed so as to not interfere with existing river outlets

[134] https://www.ovoenergy.com/guides/energy-guides/tidal-energy-what-it-is-advantages-and-disadvantages

[135] https://www.carbonbrief.org/a-rough-guide-to-tidal-lagoons/

And the general proposed arrangement for the turbines.

Images taken from BBC Newsround, 13th January 2017[136]

Plans to fully exploit the potential nationally were unveiled by TLP on 2nd March 2015. The six planned lagoons would have an expected operational lifespan of 120 years and could, they said, generate eight per cent of the UK's electricity for a total investment of £30bn. The Commons Briefing Paper (number 7940, 26th June 2018) was particularly bullish:

> ***'Around half of Europe's potential wave and tidal resource is thought to be in the UK. It has been estimated that this resource could help to meet up to 20% of the UK's current electricity demand'.)***

To a tax paying citizen/UK PLC shareholder like me, the delivery of between 8% and 20% of the country's (clean, independent, reliable) energy over a period of 120 years in return for an investment of £30bn sounds like a bargain. Especially when compared to the cost of decommissioning nuclear power plants mentioned above, or the £37bn paid

[136] https://www.bbc.co.uk/newsround/38602661

recently for the (ineffective) [COVID] Test & Trace system or £155bn[137] for HS2. Where do I sign?

Aside from the addition of significant clean, reliable and predictable 'base load' power – without the decommissioning costs and risks associated with nuclear plants – there are also other potential savings, such as on coastal (erosion) defences, and benefits, such as hydrogen generation facilities (of which more later) and other infrastructure elements, like floodwater outlets, opportunities for new recreation and leisure facilities (sailing, sailboarding, swimming, walking and cycling, cafés etc.). It ought to be possible to ensure that a large proportion of the construction was restricted to local labour, and the money, therefore, retained within the local and national economy.

Owing to cost concerns, the Government commissioned former Energy Minister Charles Hendry to review the project. The Hendry Review (published in January 2017) supported the idea of a Swansea tidal lagoon as a small pathfinder project before large-scale lagoons were rolled out, and the Welsh First Minister urged the UK Government support the project.

Despite that, on 25 June 2018, the Secretary of State (SoS), Greg Clark, made a statement to the HoC saying the project 'did not offer value for money and the Government would not enter into a contract with TLP'. He went on to say that the proposal compared unfavourably with competitors - 'on energy reliability, the generation of electricity would be

---

[137] New Civil Engineer, 14th October 2022 https://www.newcivilengineer.com/latest/dft-no-plans-to-cancel-hs2-despite-inflationary-cost-hike-claims-14-10-2022/#:~:text=Official%20costs%20for%20completion%20of,sits%20at%20%C2%A312.8bn.

variable rather than constant, with a load factor of 19% compared with around 50% for offshore wind and 90% for nuclear'. Quite where the 'load factor of 19%' comes from I do not know. Unlike waves, wind and solar, the tides are reliable and predictable for centuries to come, and there are four flows every day. If the figure was even close to correct it would represent a massive failure in engineering design.

The SoS went on to claim that 'the capital cost per unit of electricity generated each year would be three times that of the Hinkley Point C nuclear power station'. This was, of course, before the projected cost of Hinkley Point had increased by 50% and excluded decommissioning costs because the financial appraisal chose to use a horizon of just 30 years. On 5th June the First Minister of Wales had written to BEIS with a simple question: 'would you be prepared to offer Tidal Lagoon a contract of the same price and value as already awarded to Hinkley Point C?' Such a like-for-like comparison – to acknowledge the >100 year lifespan of Tidal Lagoons - would necessarily involve the inclusion of several decommissioning and rebuilding exercises for each nuclear power station (over 20bn -80% of the initial build cost – each if they follow the experience with our AGR reactors - a significant portion of costs. All in vain. The decision had been made. Similar comments apply to wind farms, which have an operational lifespan of just 20-25 years[138], but about which the SoS simply claimed that 'enough offshore wind to provide the same generation as a programme of lagoons would cost at least £31.5 billion less to build'.

[138] https://www.twi-global.com/technical-knowledge/faqs/how-long-do-wind-turbines-last

So, not even a pilot project to exploit our most plentiful, reliable power source. Perhaps the most important factor driving this kind of strategic myopia is the electoral cycle in the UK. Some issues – and national energy is one of them – require consideration on much longer time horizons than the four year maximum which that encourages.

Before we move on we need to understand what CfDs are. The government's own web site describes them as 'the government's main mechanism for supporting low-carbon electricity generation. CfDs incentivise investment in renewable energy by providing developers of projects that have high upfront costs and long operational lifetimes with direct protection from volatile wholesale prices'. So, they are the basis of 'strike prices' over given timescales, and a key device for suppliers, providing (investment return) certainty when they are planning new electricity generation initiatives, and taking care of their interests. But they also provide a basis for useful comparators which could be used far more widely as a strategic energy planning device, and which could also show value to consumers (the ultimate customers and sponsors) - using an energy-specific equivalent of a 'balanced scorecard'.

Having set aside any personal views you may have about the extent of private equity involvement, and remembering the overall goals in respect of UK energy supply, and the key attributes – most notably 'Made in Britain, for Britain', long-term Security of supply, Affordability and 'Cleanliness', we need to begin with recognition that a balanced, mixed energy supply source has been chosen as the strategy. The diagram which follows illustrates why that strategy makes sense, and the text which follows it describes how each element contributes:

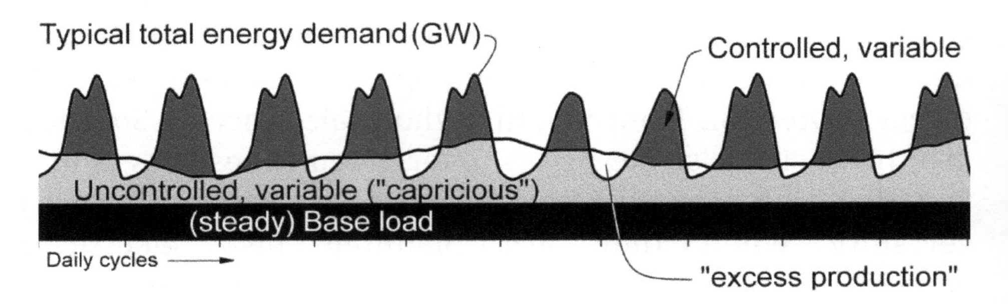

All of the data used to create this chart was made up by me, but is based on typical patterns and proportions to illustrate the points, including demand fluctuations, variations due to seasonal weather and other factors, but the general principles apply. The three different types of (electrical) energy supply considered are:

1. "Base load"

   Reliable, predictable, steady. Examples include Nuclear, Coal-fired, Hydro-power[139] and Tidal Lagoon.

---

[139] Not big in the UK, but the largest contributor of all renewable energy sources and accounts for 6.7% of worldwide electricity production https://studentenergy.org/source/hydro-power/?gclid=CjwKCAiAwomeBhBWEiwAM43YIMd6rYlVEhx6ZT3DYld5U0v0zp3OpJAB6hXVs_zMm_YLtsiKa5jnoxoC-fgQAvD_BwE

Let's assume that this represents the base value (£106/MWh in 2021, CfD price, as mentioned above)

2. Uncontrolled variable

   Suitable for supplementing base load, but capricious - can be erratic and unreliable. Examples include Wind Turbine and Solar.

   The value of this source of energy ought to be lower, reflecting its unpredictable nature, and most recently (July 2022) the UK government agreed a price of £48/MWh)[140]

3. Controlled variable

   Suitable for filling shortfalls in supply. Examples include Gas (Natural, LNG, LPG, and Hydrogen), Batteries, Hydro-power[141].

   Flexible but very expensive - current cost is estimated at £446/MWh[121].

So, the ideal model would appear to be one in which the flexible element is generated from the 'excess' produced from uncontrolled variable sources – by increasing that and/or the base load - the combination of the two ought to balance the lower and higher values of each

[140] Carbon Brief https://www.carbonbrief.org/analysis-record-low-price-for-uk-offshore-wind-is-four-times-cheaper-than-gas/

[141] In this context using pumped stored water

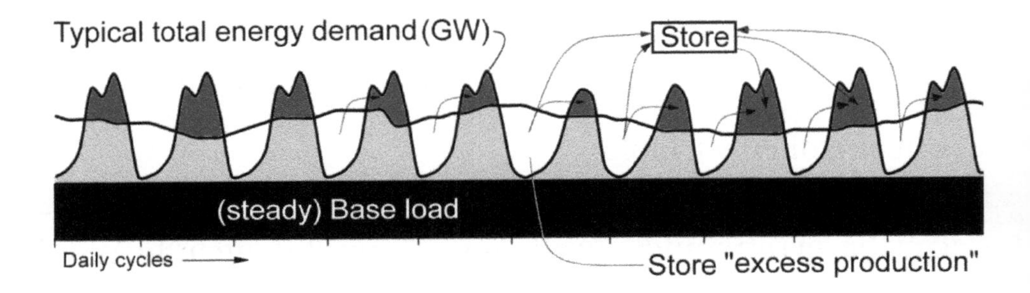

This arrangement recognises that the key element in the whole equation is the economic storage of energy, or potential energy - to enable effective asynchronous production and consumption of electricity. It could also introduce the potential for zero waste, and the development of new markets – beyond generation of electricity - to meet current demands. It could, for example, supplant fossil fuels used in other areas, such as heating, cooking and transport.

It all sounds like a good, logical plan for an optimal, clean, independent national energy strategy, promising the cheapest possible energy for UK citizens (UK PLC shareholders) and businesses. But the UK energy sector isn't currently set up like that. It is divided into three market layers – in each of which commercial enterprises operate (and each serving first the interests of their own shareholders, remember).

Generation

Many companies, from many nations, and using different generation methods

Networks

The National Grid runs the transportation (transmission and distribution) of electricity and is responsible for making sure that the supply of electricity on the grid constantly meets demand. But not everybody realises it is a private, commercial company.

Retail supply

The companies that sell electricity to consumers. There are around 60 of them, but a recognised 'big six'.

The Power Exchange (Nordpool[142] in the UK) accepts the offers of suppliers, from lowest to highest price, until demand is met, in what is known as the 'merit order'. The highest (last accepted) price is known as the 'marginal rate' (generally the fossil fuel, "controlled variable" generation), which buyers (retailers) pay to the sellers (energy generators or traders) – known as a 'pay as you clear' model. Under this model consumers always pay at the highest rate, and there is a big incentive to generators. So, in the 2022-23 winter, whilst inflation is driven by heating cost hikes which consumers struggle to afford, energy providers make record profits, regardless of effort, skill, innovation or risk.

At the time of writing there is a widespread mood of public anger targeted at power companies or their executives in relation to the record profits, whilst domestic consumers – the public – are seeing their energy costs soar. But that view is clearly misguided. Those companies are, on the whole, doing a splendid job in line with their objectives and duties. So who should we blame? Whoever set up the

---

[142] https://www.nordpoolgroup.com/en/Market-data1/GB/Auction-prices/UK/Hourly/

power supply structure in this country to serve, primarily, the interests of layers of private companies whose paramount duties and obligations are to serve their shareholders? Remember, we shareholders employ government as stewards of UK PLC. Or the people who voted for those policies (or party manifestos)?

As for the crucial energy storage, the UK has joined the global dash for batteries and electric vehicles. It feels like a move from dependence on the global market of one commodity - oil - to another - Lithium. About half of Lithium production comes from Australia, with most of the remainder coming from Chile, China and Argentina. 82% of world reserves are located in those countries, and over 70% goes into making batteries[143]. And the scramble to electric vehicles is based on the pitch that they are 'zero emission' and the premise that they are more 'environmentally friendly' than fossil fuelled cars. But, as Graham Conway explained, in his January 2020 TED talk[144] in San Antonio, the case is far from clear. The differences become far more marginal once manufacture and whole-life costs are considered. Most parts are common but, because of the complexity of battery production, the production of electric vehicles results in roughly double the production of CO^2.

[143] Statistics from the Government of Canada website https://www.nrcan.gc.ca/our-natural-resources/minerals-mining/minerals-metals-facts/lithium-facts/24009
[144] TEDxSanAntonio – YouTube: https://youtu.be/S1E8SQde5rk

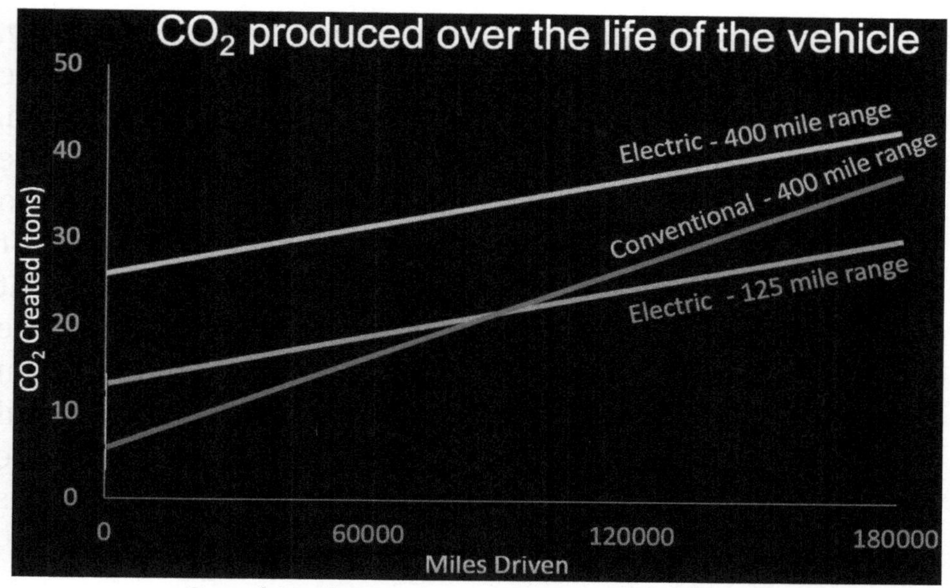

Graphic shown by Graham Conway at the TED talk

And, of course, whilst electric motors are more efficient than internal combustion engines, and nothing comes out of an exhaust pipe, there are still emissions elsewhere. The most important objective ought to be to improve our transportation efficiency. That should include maximising operational lifespans of vehicles (keep them running longer, rather than scrapping), making-do with fewer private vehicles per household, encouraging a pay-per-use model, making better use of expensive assets and decreasing the proportion of vehicles typically parked outside homes, stations and workplaces, and improving public transport.

My own view is that we should look to Hydrogen for a general national energy solution – an abundant renewable and readily available "fuel". Currently Hydrogen rightly has a bad reputation, because fossil fuels are the predominant source of industrial hydrogen - about 95% is produced by steam reforming of natural gas and other light hydrocarbons[145]. But it doesn't need to be that way. Hydrogen can also be produced by electrolysis, where electricity is used to split water into hydrogen and oxygen. This is sometimes referred to as 'Green Hydrogen'. The process is currently 70-80% efficient, but is expected to reach 82–86% before 2030. The theoretical efficiency for PEM[146] electrolysers is predicted to be up to 94%. But Green Hydrogen currently accounts for just 4% of the world's hydrogen production, because production from fossil fuels is less expensive, and fits well with the powerful petrochemical lobby. Crucially this electrolysis could be performed during periods when energy generation exceeds demand

As for use, gas-powered electricity generators could be major consumers (in place of fossil fuel gas), domestic gas central heating might be transitioned, and transport strategy could be switched from electric to hydrogen-powered cars. Many buses already run on hydrogen, and experimental work has begun on trains. In 2021 Ross Brawn, F1 Managing Director for motorsports, said ***'Maybe hydrogen is the route that Formula 1 can have where we keep the noise, we keep the emotion but we move into a different solution'***. Formula One is an area where the UK leads the world, and has a history of technical innovation, with much ending up in everyday vehicles. As Brawn said ***'we've always been a technology leader, whether that's been in safety that's come out of Formula 1 or carbon fibre materials to now how are we going to make our engines the most sustainable vehicles'***.[147] There has

---

[145] https://en.wikipedia.org/wiki/Hydrogen_production

[146] polymer electrolyte membrane cells

[147] BBC 15th July 2021 https://www.bbc.co.uk/sport/formula1/57842205

already been a brief flirtation with Hydrogen-powered cars with very limited release - primarily in Japan and the USA - of the Honda Clarity[148] - the first hydrogen fuel cell vehicle available to retail customers (2014-2021). It was voted 'World Green Car of the Year' in 2009 and 'Most important car for 100 years' by BBC's 'Top Gear programme. It took less than 5 minutes to refuel, and had a range of 366 miles but the car itself didn't solve the infrastructure shortcomings - a lack of Hydrogen filling stations, and availability of Green Hydrogen – so it failed to break into the general market.

Green Hydrogen seems to offer a key opportunity for the UK to be a world leader, whilst offering domestic security of clean energy availability. Imagine having a national strategy which targets development of Green Hydrogen generation and use; a Green Hydrogen economy. Perhaps even a collaboration programme with F1 and Honda. Personally I would aim to build the generating plants into, or alongside, the new Tidal Lagoons and where offshore wind farm power comes ashore. But then I'm not a Civil Engineer. My Father was.

---

[148] Wikipedia https://en.wikipedia.org/wiki/Honda_Clarity

Outsourcing/privatisation

If you can set aside political ideology, you might be questioning whether essential utilities with local monopolies are suitable candidates for privatisation. In Britain we have privatised most utilities, and often rely on the oversight of separate bodies – usually in the public sector – to monitor and "govern" them. Despite this, in many areas the public are beginning to realise that outsourcing arrangements can be and are "sweetened" for suppliers by loosening of standards and regulations or by hobbling of oversight bodies.

In my humble opinion, whilst commercial competition in fields of discretionary spend can be a healthy driver of innovation and value, where a monopoly service is critical to the proper functioning of the country, and would not be allowed to cease if they fail – like water or power supply, waste disposal, or communications infrastructure – then it is not suitable for outsourcing. The first responsibility of any government is to take care of its citizens, and it is not possible to 'outsource' that responsibility. And yet in Britain we have done just that with privatisation of most utilities.

Very few of Britain's industries were publicly owned prior to the Second World War. The General Post Office (GPO) had been around since the middle of the seventeenth century for post, with licensing of wireless telegraphy from 1870, of (previously largely private) telephone networks from 1880, and radio from 1904. In 1922 the private, GPO-licensed British Broadcasting Company began – replaced by the British Broadcasting Corporation (BBC) in 1927. The National (electricity) Grid was established in 1926 by the Central Electricity Board, and London Transport was formed in 1933.

In 1945 Clement Attlee's Labour party were swept to power and quickly embarked on bringing into public ownership key industries, as promised in their manifesto. That incredibly ambitious body of work included the creation of the National Health Service and a welfare state that offered unemployment benefits and retirement pensions in return for standard 'National Insurance' payments, the nationalisation of the railways, coal mining, the iron and steel industries, gas and electricity. England and Wales' water industry - supply, sewage and river management – was nationalised later, in 1973.

Following the Conservative Party's election victory in 1979, Margaret Thatcher reversed much of this work during the 1980's, but she began by first planning and commencing the weakening of Trades Unions in order to make the businesses more attractive to investors, and by selling off council houses at a discount in order to free up public equity and create a new class of equity-rich capitalists hungry to invest. It was in her second term, after winning the 1983 election with a huge 144 seat majority following victory in the Falklands war that, alongside increasing those measures, she set about the wholesale privatisation of public assets.

British Airports Authority, Associated British Ports, Trustee Savings Bank, British Gas, Royal Ordnance, British Airways, British Rail Engineering, British Shipbuilders, British Steel,

British Sugar, British Telecom, Britoil, Enterprise Oil, Cable and Wireless, Municipal bus companies, National Bus Company, National Express, National Freight Corporation, Harland and Wolff shipbuilders, Water companies, Unipart, and a large number of major engineering companies, including British Aerospace, Rolls-Royce, British Airways Helicopters, Jaguar, British Leyland, Rover Group, Leyland Bus, Alvis, Leyland Trucks, Leyland Tractors, Fairey, and Coventry Climax, and technology companies, including Amersham International, ISTEL, Ferranti, and Inmos were all sold off. And, of course, many more Council houses at discount rates under a 'Right to buy' scheme.

At the time I was naive (and even poorer, so I was spared any moral discomfort!) but witnessed the unseemly scrabble amongst more senior colleagues at the bank where I worked at the time, to buy as many of the publicly released shares in utilities as possible. 'It's a licence to print money' they said, 'a risk-free investment'. They were right, of course – the utilities could never be allowed to fail, and share prices were discounted to sell.

Between 1989 and 1990, companies privatized by the Thatcher government fattened the government purse by some £2 billion[149].

In 1996 the MOD's 55,000 Married Quarters estate was sold to a consortium led by Nomura Bank of Japan for £1.6bn[150]. Guy Hands' (who used to work for Nomura) 'Terra Firma'

---

[149] Harvard Business Review, January – February 1992 https://hbr.org/1992/01/british-privatization-taking-capitalism-to-the-people

[150] https://www.heraldscotland.com/news/12033898.old-soldiers-disgusted-as-japanese-win-property-deal-outcry-at-mod-homes-sell-off/

bought the estate from Nomura for £3.2bn in 2012[151] (both Hands and Terra Firma are based in the tax-haven of Guernsey). The National Audit Office later found, in their 2018 report[152] HC762 that the deal left us £2.2bn–£4.2bn poorer than had the MOD retained the estate.

Worse was to come. In 2001, using a PFI[153] deal, the Inland Revenue sold more than 600 of its buildings to a Bermuda-based company called Mapeley Steps Limited - part of Bermuda Mapeley Holdings Limited, a company ultimately owned by George Soros and US group Fortress Investment, for £220m. Truly irony was dead - the people who worked on our behalf to ensure that UK tax was paid now operated from premises owned by and leased from a tax haven-based company. I recall that the mainstream press were silent about the matter at the time, although Private Eye magazine diligently reported on it, and it wasn't until over a year later that the Inland Revenue admitted their 'error' and the matter was more generally reported[154].

There have been fewer privatisations since, but a notable example was Royal Mail. They seemed to have managed perfectly well in public hands for 499 years, but were privatised in 2011/12 'to comply with EU legislation'. Of course as leading members of the EU we will have had a major hand in drafting and agreeing that legislation, and it's odd that, for example, the French state still owns La Poste and that over two years since we left the EU in order to 'take back control' Royal Mail still hasn't been taken back

[151] https://www.independent.co.uk/hei-fi/business/guy-hands-swoops-for-army-homes-in-ps3-2bn-deal-8334965.html
[152] https://www.nao.org.uk/wp-content/uploads/2018/01/The-Ministry-of-Defences-arrangement-with-Annington-Property-Limited.pdf
[153] Private Finance Initiative
[154] The BBC http://news.bbc.co.uk/1/hi/business/2263208.stm

into public ownership. Coincidentally, over £2bn has been paid out to shareholders of Royal Mail since privatisation.

Revered institutions like the NHS and social care have proven to be tougher nuts to conquer, but they seem to be in the crosshairs. As Noam Chomsky said, in his lecture 'The State-Corporate Complex: A Threat to Freedom and Survival', given at the University of Toronto, 7th April 2011 – 'there is a standard technique of privatization, namely defund what you want to privatize. First thing to do is defund them, then they don't work and people get angry and they want a change. You say okay, privatize them and then they get worse.'

Ian Byrne MP summed up the position succinctly in his 4th December 2022 Tweet: '1970s Britain owned its water, transport, mail, energy, millions of council houses and sat on a huge North Sea reserve. The national debt was £80bn. 2022 Britain has no assets and a national debt of £2.5Tn'[155].

The water industry in the UK stands as a shining example of what decades of failing to govern in the interests of citizen stakeholders can lead to, and several consequences can be illustrated through it.

---

[155] At the time of writing it's over £2.7Tn https://www.nationaldebtclock.co.uk/

UK water

I do not know of a single other major country that has privatised its water supply. If there is any utility which should never be allowed to fail, it is surely this, and therefore entirely unsuitable for privatisation. When the ten publicly-owned water companies were privatised in 1989 it was meant to usher in a new era of efficiency and investment – to draw in private capital to fund repair and upgrading of ageing infrastructure, to improve service and lower customers' bills. But that is not how it has turned out. My colleagues, who greedily gobbled up the shares, recognised the nod and understood the lack of risk and promise of rewards. More than 2.5 million people applied for shares, and the offer was nearly six times oversubscribed. The average gain to investors on the first day of trading was 40% and over the next two decades, rather than injecting funds, the privatised water companies paid more than £57bn in dividends, at the same time as running up large debts, the interest on which is effectively paid for by customers[156].

Fast forward a few years and we have arrived at a position where, as taxpayer citizen UK PLC Shareholders we are typically paying a private (often offshore) company to release our own poo into our rivers and onto our coastlines, whilst our politicians cover their backs by reducing regulation and penalties and starving regulators of funding. Take my local water company as an example. They were

[156] https://www.theguardian.com/commentisfree/2022/aug/16/i-worked-on-privatisation-england-water-1989-failed-regime

recently fined for pumping raw sewage into the sea. Between 16bn and 21bn litres (equivalent to 7,400 Olympic-sized swimming pools) of raw sewage, in 51 (admitted) cases covering 6,971 (out of 8,400) illegal discharges over nearly six years. I have swum in that sea many times over the period in question, and shall need to again, since our local public swimming pool has recently been closed due to the need to make budget savings, so I have reason to care more than most.

The company[157], which is owned offshore (presumably to avoid the need to pay UK tax), was fined a record £90m which, the judge said, 'should be a deterrent to other companies[158] and might act to prompt shareholders to ensure that the utility improved its regulatory compliance'. This seems unlikely, because it followed a history of criminal activity with its 'previous and persistent pollution of the environment' (it had 168 previous offences and cautions but had ignored these and had not altered its behaviour). The dumping also enabled Southern Water to avoid penalties of more than £90m, according to OFWAT[159], as a result of failing to meet strict standards on discharging wastewater. Essentially, it was (and remains) less expensive for the company to pay the fines than to properly treat sewage. Unsurprisingly, within 72 hours of the verdict there were over a dozen reported[160] new 'releases' by them.

Meanwhile Southern Water claim, in their Annual Report and Financial Statements for the year ended 31 March

[157] Southernwater.co.uk. The ultimate parent company and ultimate controlling party is Greensands Holdings Limited (GSH), a company incorporated in Jersey
[158] There are about ten
[159] The water sector regulator in England and Wales
[160] https://www.southernwater.co.uk/water-for-life/our-bathing-waters/beachbuoy

2020, that their 'purpose and strategic priorities' are to 'provide water for life to enhance health and wellbeing, protect and improve the environment, and sustain the economy, in order to create a resilient water future for our customers'. But remember, the first, and primary duty of a Limited company and its executive officers is to maximise the value of the company and return to its shareholders.

The company made an operating profit of over £328m in the year to 31/03/2020, and its highest paid director received aggregated emoluments and benefits of over £1m.

The fine, by the way, goes direct to HM Treasury, whilst the Environment Agency[161] - which is left to clean up or deal with the consequences of the mess - has had its budget cut from £120m to £40m.

Southern water is not an anomaly. The London Evening Standard reported[162] on 21st January 2022 that 3 billion litres of untreated sewage had been dumped into the River Thames during 2020 (a rise of 600% from 2016), two billion litres of which was in just two days (during a 48-hour period in October) (enough to fill 800 Olympic-sized swimming pools).

It is, of course, much cheaper for the water companies to simply release untreated sewage than to treat it first, so better suited to their essential goal of maximising profits for their shareholders.

---

[161] An executive non-departmental public body, sponsored by the Department for Environment, Food & Rural Affairs, whose stated remit is to 'work to create better places for people and wildlife, and support sustainable development'.

[162] https://www.standard.co.uk/news/london/two-billion-litres-sewage-dumped-into-river-thames-water-b977981.html

In 2021 there were 372,533 sewage spills, over a period of 2.7 million hours. The position was worsened in September 2021 when the EA introduced a waiver[163] allowing some companies to not have to go through the third stage in the treatment of sewage if 'supply chain failure' (a.k.a. Brexit) meant that they did not have the right chemicals. Seven weeks later, following a dogged campaign led by various environmental organisations, Swimmers Against Sewage and Feargal Sharkey (the former 'Undertones' singer), an amendment to the Environment Bill to place a legal duty on water companies not to pump waste into rivers was brought (by the Duke of Wellington) in the House of Lords (HoL). But the Government took the side of the water companies' shareholders rather than that of its own citizen shareholders and the environment of the UK and used its 80-seat parliamentary majority to vote it down.

A report[164] published on 13th January 2022 by Parliament's Environmental Audit Committee (EAC) stated (s133) that 'Without these overflows, sewage could potentially back up into domestic and commercial properties when the sewerage system is overloaded, for instance in periods of heavy rainfall. Overflows are intended to be used infrequently and under exceptional conditions: this is reflected in the permit conditions stipulated by the Environment Agency. Their use nevertheless appears to be increasingly routine, as pressures on the sewerage network grow. Monitoring data seems to show instances where

[163] https://www.gov.uk/government/publications/water-and-sewerage-company-effluent-discharges-supply-chain-failure-rps-b2?utm_medium=email&utm_campaign=govuk-notifications&utm_source=888a69f8-3f78-4fee-bc00-d6b39eea1e8b&utm_content=daily

[164] https://committees.parliament.uk/publications/8460/documents/88412/default/

overflows are being triggered at times of low or no rainfall'. 'The number of sewage spills from overflows officially recorded by water companies and reported to the Environment Agency (EA) reached 403,171 in 2020, a 27% increase on the 292,864 recorded in 2019'.

On 25th January 2023 the government, again leveraging its huge majority, finally defeated HoL opposition and granted water companies a further 15 years of impunity through a new statutory instrument (SI), whilst setting targets to reduce water pollution by 80%.

The level of public outrage directed towards the water companies baffles me. They are simply doing what they exist to do – maximise the return to their shareholders – and seem to be doing it very well, balancing costs and benefits, assessing net consequences of each of their options, and in the absence of market competition. I share the anger, but I wish it would be directed in the right direction – towards those who transferred control of our water supplies and waste disposal in <u>our</u> interests to private companies whose paramount duties and obligations are to serve their own shareholders, whilst simultaneously disempowering and defunding those whom we employ, as stewards for UK PLC to regulate them. .

The current status was well described, I think, by Phillip Oppenheim - Conservative Exchequer Secretary to the Treasury between 1996 and 1997 – in a letter to the Financial Times in September 2022:

> ***'If financial services produced wealth for the whole economy, the UK would be a stunning success, not a country which has sunk from sixth in per capita gross domestic product in 1960 to just 20th now***

(excluding tax havens and oil rich nations, but including Norway). This puts the UK well below the western European average.

While the dollar/sterling exchange rate grabs the headlines, the pound now lies 15 per cent below its 20-year average against the euro, the currency derided by many Conservative politicians. We are even hurtling towards parity with the Swiss franc - in 2000 the rate was 2.5 to the pound.

Overfinancialisation of our economy and misallocation of resources is one cause of low UK investment and poor productivity.

Rather than reducing regulation and tax, we should be ending the carried interest tax break for private equity; better regulating pensions and fund managers who top-Slice their percentage from our pensions, regardless of performance; preventing, rather than encouraging pension funds from investing in risky and opaque PE; and stopping PE from treating utilities like milch cows and offshoring products to tax havens.

Nothing in last week's mini-Budget indicates that our new leaders have the slightest grasp of the scale of our long-term, structural problems - or the solutions, above a half-digested, two-dimensional version of Thatcherism.

Rather than blustering about Britain being the "fifth richest economy", we need honest leaders who accept that we are no longer a world power, stop pretending there is a special place for us in the anglosphere or Commonwealth and

concentrate instead on internal development. We need to rein back the worst abuses of financial services; reform education; encourage savings over consumption; reduce incentives to invest in property rather than productive assets; Smarter, not higher defence spending - no more aircraft carriers; and, of course, rejoin the EU single market'

The House of Lords

Since the subject came up in the previous section I thought we'd take a brief look at the House of Lords (HoL). It's currently a hotbed of political dogma, but what is really needed is a dose of cool, considered logic.

The HoL – the 'Upper Chamber' of England's legislature – has its roots back in the 13th or 14th century, and was founded in its current form in 1801 following the Acts of Union with Scotland (1707) and Ireland (1800) as a single parliament for Great Britain and then for the United Kingdom. Its function is to scrutinise bills that have been approved by and to act as a check on the more powerful House of Commons (HoC). Unlike the HoC[165] it does not have a fixed number of members, and they are not elected by members of the public. They are supposed to offer a degree of stable expertise, experience and wisdom. From an initial cohort of around 50 the number has grown enormously as, alongside periodic suggestion of reform, governments of various persuasions have dished out peerages in return for party loyalty or donations. The current number of members is 760[166] including 92 hereditary peers who are there because an ancestor won a battle, bought a peerage or somehow pleased or was particularly useful to somebody in power. And 26 Bishops.

The HoL is currently facing a barrage of calls for reform, ranging from disbanding, through democratic election of members and all points in between, with a host of tinkering

[165] Which currently has 650 members

[166] The second-largest legislative chamber in the world. Only the Chinese National People's Congress has more.

ideas, including potential relocation. It's difficult to argue that reform isn't needed, but it's less easy to agree what form that should take. In my humble opinion, reform would be less contentious if focused on purpose - by focusing not on the democratic function - which is already performed by the HoC – nor patronage, but on helping it to better perform its vital role of expert review and quality improvement in the government of Britain.

I'd like to see an end to the notion that seats are gifted to loyal friends and/or donors and for all party politics to be taken out of the Upper house by removing any party whip from members. The House would then be free to concentrate on its true purpose and function in a practical, fact-based manner in the national interest, free from political dogma or influence. And I'd also like to see it slimmed-down, commensurate with that role.

Why not have the second, 'expert' chamber comprised of members elected by existing representative bodies? There are usually professional bodies or mechanisms which could handle the election of the person(s) to represent their field – or their own advocate. The only challenges would be to agree what the appropriate total number of members should be, which fields ought to be represented, which bodies should be the electors, and the term of office. Each of the members would be elected by an appropriate professional body, trade association or similar to serve for a fixed term – say two years - with a deputy to cover during periods of absence.

For what it's worth, if I were asked, my initial suggestion to start the ball rolling would include;

> ***doctors, nurses, teachers, engineers, labourers, scientists, shop workers, parents, sole trader***

electricians, plumbers, carpenters and decorators, students, religion representatives (say the top three or four religions[167]), manufacturing operatives, motor mechanics, cyclists, musicians, artists, farmers, pensioners, sports(wo)men and so on.

Agreeing that list of interests and expertise would probably be the most challenging task of all, and might require modification after an initial trial period but would, otherwise, be a one-off exercise. Changes should be subject to a rigorous regime akin to that governed by the Boundaries Commissions[168]. Initially analysis of the numbers of people who would be represented might be helpful, and a recognition that specialist representation cannot anticipate every issue which might arise. Each primary representation member (the title probably needs some work) would be eligible to take on advocacy additionally on behalf of a secondary interest. That interest might be something close to the representative's heart, like a hobby, or be a sub-set of their primary interest. Say beekeeping, or philately. Every representative will likely have multiple experiences, knowledge and expertise - some teachers may also be fathers, some engineers may also be mothers and car drivers, and some students may also use public transport, for examples - and every special interest group would likely seek them out and be keen to secure a formal advocacy of that interest. Ultimately every member of the house would have one primary representation mandate and a number of secondary ones. Paid advocacy

---

[167] currently in the UK, in percentage order: Christianity (Church of England, Catholic, others), Irreligion (Atheist, Humanist), Islam, Hinduism - https://en.wikipedia.org/wiki/Religion_in_the_United_Kingdom

[168] The Boundary Commissions are independent and impartial non-departmental public bodies, which are responsible for reviewing Parliamentary constituency boundaries. There is a separate one for each of the separate countries of the UK.

should, of course, be strictly prohibited, so appointment from industry would probably need to be on the basis of a secondment, end of career, or similar.

There is also lively debate about reform of the HoC, whose 650 members are elected democratically based on geographical constituencies. That debate centres principally around proportional representation. Anyone who has watched Prime Minister's Questions in the HoC cannot be anything but ashamed that the spectacle is broadcast to the world. Watching 'the Mother of Parliaments', governing 'the fifth largest economy in the world', performing some arcane pantomime in the 21st Century can be toe-curlingly embarrassing. Although I think that a greater degree of constructive, cross-party collaboration in the HoC for the national good would be welcome, I don't intend venturing into that minefield here. But we ought to recognise that if the impartiality and caliber of reviewers in the upper chamber was improved then it would carry greater respect and authority, resulting in improved consideration of issues by and output from the HoC, to the benefit of the national governance mechanism as a whole.

Taxi for Mr Swan

'Don't it always seem to go, that you don't know what you've got 'til it's gone?' - Joni Mitchell, 'Big yellow taxi'

This is not a crafty way of sneaking a bit of autobiography into the book (my life is far too boring and inconsequential), but the next few pages give essential context for what follows.

I didn't expect that

'We must be willing to let go of the life we planned so as to have the life that is waiting for us' – Joseph Campbell

Wednesday 31st May 2017 promised to be an unremarkable day. We had recently moved into a new house in order to care for my Wife's elderly mother. I'd spent a year during evenings after work and at weekends working and overseeing specialist contractors to renovate and modify the house for the purpose. I had retired a month previously from full time work (and commuting) in readiness for the demanding task of helping to care for someone suffering from advancing dementia.

My Wife and I woke at about 7am and began dressing in readiness for taking the dog for a walk. I was struggling to put on my trousers. 'I think I've had a stroke in the night' I said, intending to make light of my clumsiness, rather than delivering a clinical diagnosis. 'Just hurry up and get dressed' said my Wife, who'd had over forty years' experience of my "sense of humour". We drove the four miles to a favourite spot; peaceful and with plenty of space for me to kick Bertie's tennis ball for him safely. It was a fine spring day, and a welcome hour's respite before a day of new challenges, as we learned how to best fulfil our new carer roles. My boots felt a little heavier than usual as we climbed the gentle hill on the last part of the walk, though the drive home was uneventful and filled with normal conversation about the day ahead. And when I clumsily pressed the accelerator along with the brake when parking on our driveway, I put it down to the clunky walking boots that I was wearing. But when I came to get out of the car I discovery that I couldn't stand up. How peculiar! My Wife, who had already got out of the car, brought her mother's walking stick from the house, but I couldn't even hold it properly. More worryingly I was apparently talking even more gibberish than usual. My Wife helped me to shuffle into the house, and then sat me on a sofa before calling 999. I began shaking. I couldn't seem to speak coherently nor control my body. I was scared and confused. An ambulance arrived in about two minutes (luckily it had been just around the corner at the time) and two paramedics then questioned me and checked me over. Having established that I had probably suffered a stroke, they loaded me onto a chair and into the ambulance (I remember feeling guilty that I couldn't help, and having my first taste of being entirely powerless and trusting the capable hands of others). During the a 20-mile blue-light trip to hospital, I had what I imagine is like a panic attack. Whilst the paramedics tried to engage me in conversation and keep me conscious, my body was in full survival mode; sweat was pouring from every pore, I couldn't speak, couldn't breathe properly and wasn't even trying to put on a brave face any more. All that I remember about the handover at A&E was that I wasn't able to properly express my gratitude to the paramedics who had kept me going and got me safely to the hospital. I can't remember the last time I was in hospital as a patient, but I really did just succumb – confused and scared, but comfortable. It seemed strangely reassuring to be divested of my watch, money and belt by people who seem to have done this a hundred times before, and in whose expert hands I was happy to be. I don't remember anything between being fed into an MRI scanner and half waking in an isolation room. I shall probably never know the extent to which they were drug-induced or the result of my damaged brain figuring out what had happened to it, but that night I fell into a deep sleep filled with vivid and confusing dreams.

Had it all just been a bad dream? Apparently not. The following day I woke to find that my right arm had been replaced by a hanging lump of meat which, although it looked like my arm and hand, I could no longer operate. It had become luggage, and my right leg was almost as useless.

Throughout the experience – from the initial stroke itself, to the initial intensive care – there was no physical pain. Plenty of discomfort, anguish, frustration, but no pain. I later found the documentary film 'My Beautiful Broken Brain' (2014) by Lotje Sodderland about the stroke which she suffered herself at the age of 34 in 2011 to be an excellent reflection of some of the experiences and emotions that I went through (though fortunately less severely). It has its lighter moments, and I can recommend it, (although it is only available on Netflix).

It became all the more real for me when my family arrived all doing a poor job of concealing the fact that they'd had prior waiting room discussions of concern. I also discovered that I had lost the power of coherent speech. Needing to ask a nurse to bring me a 'bottle' in order to have a wee was a new experience, but at least it proved I still had control of that essential body function. I lost a lot of weight over the next week or so because I barely ate (how DO you sit up and operate cutlery when you have only one working hand?) and feared the need to find a place and method to go for a poo (it was several days before I went). All modesty and personal vanity needs to quickly take a back seat when there are more pressing issues whilst you're in a hospital Intensive Care Unit (ICU) and, anyway, most others there are in a similar position.

By the time the Consultant Neurologist came to see me I had been moved from an isolation room into the ICU ward with others, suggesting that I was out of the woods, and he was reassuringly upbeat and optimistic. Apparently the brain damage shown up by the scans was relatively minor, but, as a naïve newcomer to this kind of thing, and still in a state of shock and fear at my sudden impairment, I asked him for his prognosis - would I recover fully, and how soon? I was so heartened by his confident prediction, delivered with a reassuring smile ***'You should make a 96% recovery'*** that I don't recall asking him to clarify what that meant, and how quickly it might be expected? He did tell me that being right handed was helpful, because it should speed recovery.

I should've expected that

'There is only one kind of shock worse than the totally unexpected: the expected for which one has refused to prepare' —-Mary Renault, The Charioteer

The possibility of suffering from a stroke hadn't previously even crossed my mind. A heart attack, cancer, or being wiped-out on the road, maybe, but who has a stroke?

It turns out that they are quite commonplace; someone in the UK suffers a stroke every 5 minutes, and despite only about a third dying it's still the fourth biggest killer. More than half of those who do survive are left disabled and dependant on care by others[169]. There are 1.3 million stroke survivors in Britain[170] and, for some reason, the number is increasing. And now I think about it, my Brother, Cousin, Aunt and Grandmother had all suffered strokes.

At the time of my stroke I was 61 years old, reasonably fit and was looking forward to retirement, and time to pursue a number of activities after 44 years of predominantly desk-work. I had given up smoking more than 15 years previously and drank very little – maybe a pint or two of bitter per week at most. I had already been offered a couple of consultancy commissions, which I thought would keep me mentally active, and to which I could add to fill available time when interesting opportunities arose.

---

[169] Stroke Association, 2008

[170] Stroke Association 2023

Stress had been a constant factor throughout my four decade working life, and the commuting routine - 04.15am wake-up alarm, a 5½h/day £6.5K/year commute via unreliable rail services and no certainty about arrival at either end, let alone times - had unquestionably taken its toll.

Then a combination of circumstances had begun to result in an increase in my blood pressure. A Cardiologist gave an initial diagnosis and said that she would arrange for an echocardiogram in order to verify her diagnosis and help her to determine appropriate treatment. But the NHS had already been subject to 'Austerity' cutbacks for several years, and I was still waiting for the scan three and a half months later when I suffered the stroke.

At work numerous organisational changes including a major shift in operating model and a corporate merger, during which I had three different managers within a year and a total lack of direction, didn't help. Early in March a close colleague had a stroke, just as he was retiring, which I took as a final sign and decided to retire myself, giving notice that same month.

We all have 20-20 Hindsight and I can now see the signals that were missed. A couple of months earlier I had had a couple of "funny (dizzy) turns" – one when I had been driving and needed to stop, then episodes of flashing lights around my peripheral vision, that I guessed were migraines, which I had never experienced previously, and a couple of days of excruciating headaches which felt as though they were from behind my left eye. Headaches were extremely rare for me, and usually quickly resolved by drinking a glass of water. Finally I had a sudden partial loss of hearing, which began suddenly during a particular moment of stress. It was so strange that I consulted my doctor, who peered into my ears and declared them fine, although my hearing didn't recover fully before the stroke.

But hey, this was no worse than I had endured for years. I was bullet-proof, and looking forward to a stress-free retirement full of country walks, painting, tennis, carpentry and quality time with my family. Unfortunately I didn't have the time to do much of any those things during the month that I had between retirement and the stroke.

96%

'Life can only be understood backwards; but it must be lived forwards' - Soren Kierkegaard

The 4% deficit forecast by my neurosurgeon sounded pretty good at the time. I felt as though I had dodged a bullet. After all, over half of British voters enthusiastically voted for Brexit[171] and the 4% hit to the UK's GDP, (as accurately predicted by the Government, experts, and the OBR[172]. That's about £80,000m[173], or more than we spend on education (£76.4bn 2021-22), almost twice what we spend on defence (£46bn 2021-22)[174] and double the effect on the UK economy of the global COVID-19 pandemic.

But let's not forget that chimpanzees share at least 96% of their DNA with humans according to the National Geographic[175].

I decided that I'd just knuckle-down (is that a Freudian chimpanzee reference?), do whatever the clinical experts recommend, and anything else I could, and 'take my '4%' medicine'. It could be worse. I know that I am lucky; many don't survive, and of those that do many suffer much greater impairment than me. But, 4 years on, I can't walk

[171] Britain's exit from the European Union (EU), in January 2020, following a referendum in 2016. https://obr.uk/box/the-initial-impact-of-brexit-on-uk-trade-with-the-eu/

[172] Office of Budget Responsibility - created in 2010 to provide independent (of Government) and authoritative analysis of the UK's public finances.

[173] https://www.ons.gov.uk/economy/grossdomesticproductgdp

[174] https://obr.uk/forecasts-in-depth/brief-guides-and-explainers/public-finances/

[175] https://www.nationalgeographic.com/science/article/chimps-humans-96-percent-the-same-gene-study-finds (other sources suggest an even higher percentage)

properly or far, speak coherently for long, or do most of the recreational things which I previously loved. And I'm typing this with one finger, slowly.

No two strokes are alike - at least in terms the way they affect the victim - and everyone who suffers from one has a different background. But some aspects are common, so I offer my own experience only as an illustration, in the hope that some may resonate with and be helpful to others.

I had been told that mine was an ischemic stroke (a blockage of blood flow to the brain, resulting in some tissue dying) in the left hand side of my brain, affecting, principally, my right limbs, fine-motor control, and speech. I was also already familiar with the term 'brain plasticity'. But this had suddenly become something interesting and important in my life, and knowing more might speed and improve the extent of my recovery. So I embarked on a new 'project'. First challenge was mobility – regaining use of my right arm and leg, and speech, so that I could move around, use the toilet, eat independently, and communicate. A Zimmer[176] frame became my new best buddy. Not only could I use it to navigate from my hospital bed to the ward toilet, but it was also somewhere to rest my (useless) right hand. It also allowed me to shuffle around the hospital corridors in the vain hope of regaining the ability to walk, and, with other stroke victims, shuffle down to the therapy room, where Occupational Therapists presented us with manipulation challenges/tasks to try to re-form the connections between brain and body in the vital few days after a stroke. It's difficult to remember, let alone explain, the frustration of trying to roll a squishy ball on a desktop

[176] Interestingly Zimmer now make all kinds of assistive medical kit, but not that which will forever bear its name

from my left to right hand. I guess the plan was for memory to take over but all I can remember is staring at 'my' hand and trying to will it to obey me. It stubbornly refused. At night I adopted a foetal position with my left hand holding its right sibling, in the hope that it might trigger memory overnight. It didn't, but at least it kept the rogue limb out of harm's way.

Food in the hospital was excellent and healthy, but I discovered that knives and forks are surprisingly tricky blighters when you only have one hand to work with, and that one is the 'wrong' hand. And those little foil-wrapped butter-pats and salt and pepper sachets might as well be bullion safes. Being washed on a stool in the shower by an unknown young lady was also a first for me. Gosh, our nurses do have a lot to contend with, and they do it with such diligence and grace.

After about a week of having my vital functions monitored 24/7 I was discharged to begin my longer-term recovery journey at home. I was accompanied by a party bag full of drugs which would now be a part of my life and, still, the alien useless arm & hand, which I hoped would not. I'm not sure I had expected to see home again, but it was great to be there, despite all of the new challenges, like stairs, sleeping without the constant sound of beeping, and getting dressed.

Once the initial shock had subsided, and I had settled back in at home, I was eager to crack on with my 96% recovery. I was lucky to have support from the local NHS Occupational Therapy (OT) team for a few weeks, initial support and welcome advice from the Stroke Association, and the loving support of my family. I described to the OT the overall effect as being like ageing 20 years within an hour on that fateful morning, and I later created a graphic to illustrate the point.

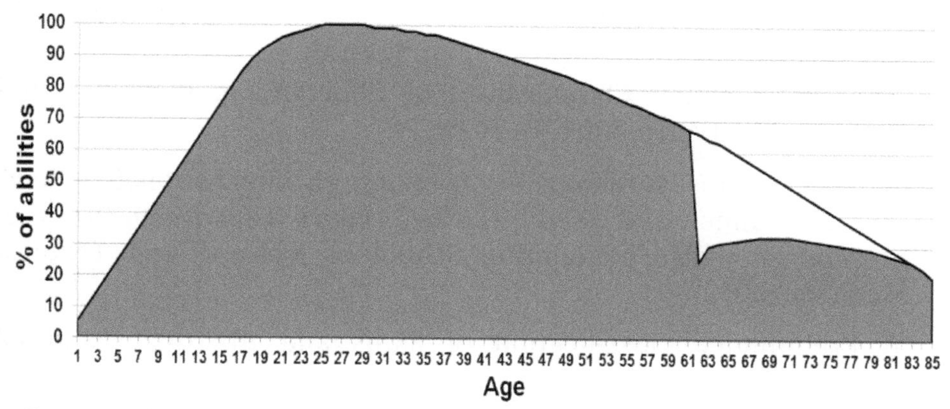

Human abilities over a lifespan – showing the effect of a stroke on mine

The chart is intended to show the typical profile of the "abilities" of any given individual and an average lifespan of about 80 years.

My own profile – illustrating the effect of the stroke at the age of 61 – is overlaid, and illustrates the "loss" which I perceive. After initial recovery most is damage-limitation work-arounds. Others might usefully be able to plot their own "abilities profile" to reflect significant changes – both good and bad – over the course of their own lifetime.

With my Business Architecture hat on I sought out an existing model of the kind of "abilities" that I had in mind – ranging from feeding oneself, through toddling, peak athletic and mental prowess, and later fragility and infirmity. I considered

> 'Taxonomies of Human Performance: The Description of Human Tasks', January 1984

> 'Development of a Taxonomy of Human Performance: A Review of Classificatory Systems Relating to Tasks and Performance, Technical Report 1', George R Wheaton, December 1968

'Falls Efficacy Scale', Tinetti, M., D. Richman, et al. (1990)

'Proving disability and reasonable adjustments A worker's guide to evidence under the Equality Act 2010' edition 5, by Tamara Lewis for Central London Law Centre

'The International Classification of Functioning, Disability and Health (ICF)', (WHO 2001:5), 2018

'Models and measurement in disability: an international review', Michael Palmer and David Harley, Oxford University Press in association with The London School of Hygiene and Tropical Medicine, 2011

'A Preliminary Classification of Skills and Abilities', National Research Council, 2012

There is also the 'Critical Frailty Scale', used by the NHS for the over 65s

1 Very Fit – People who are robust, active, energetic and motivated. These people commonly exercise regularly. They are among the fittest for their age.

2 Well – People who have no active disease symptoms but are less fit than category 1. Often, they exercise or are very active occasionally, e.g. seasonally.

3 Managing Well – People whose medical problems are well controlled, but are not regularly active beyond routine walking.

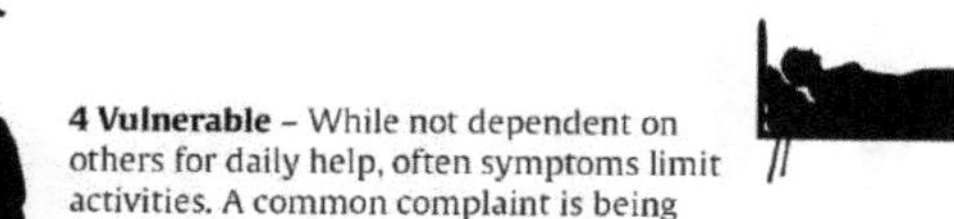

4 Vulnerable – While not dependent on others for daily help, often symptoms limit activities. A common complaint is being "slowed up", and/or being tired during the day.

5 Mildly Frail – These people often have more evident slowing, and need help in high order IADLs (finances, transportation, heavy housework, medications). Typically, mild frailty progressively impairs shopping and walking outside alone, meal preparation and housework.

6 Moderately Frail – People need help with all outside activities and with keeping house. Inside, they often have problems with stairs and need help with bathing and might need minimal assistance (cuing, standby) with dressing.

7 Severely Frail – Completely dependent for personal care, from whatever cause (physical or cognitive). Even so, they seem stable and not at high risk of dying (within ~ 6 months).

8 Very Severely Frail – Completely dependent, approaching the end of life. Typically, they could not recover even from a minor illness.

9 Terminally Ill – Approaching the end of life. This category applies to people with a life expectancy <6 months, who are not otherwise evidently frail.

Clinical Frailty Scale[177]

All fine bodies of works, but not what I had in mind. The nearest, conceptually, that I found was in a 1965 book called 'The structure of human abilities' by Philip Vernon, but even that proved to be less helpful than I thought the name implied. I reckon that if Mr. Vernon took a whole book to explain his theory then I was unlikely to be able to do justice to mine in a paragraph or even a chapter, and, anyway, most people that I've shared it with seem to readily understand the concept behind mine. I think we all know, instinctively, the pattern of life; the enthusiasm and naiveté of youth when every day brings a new peak of

[177] The Rockwood Clinical Frailty Scale, developed by Ken Rockwood at the Dalhousie University, Nova Scotia, and used by The Royal College of Physicians and included in NHS NICE Critical Care Guidelines *https://www.criticalcarenice.org.uk/frailty*

achievement, through our 'prime', to the points when we become less good at or unable to do things, and that not all injuries will fully heal. But we tend not to talk about it, unless in terms of trite expressions such as 'live each day as if it's your last'.

Cradle to grave[178]

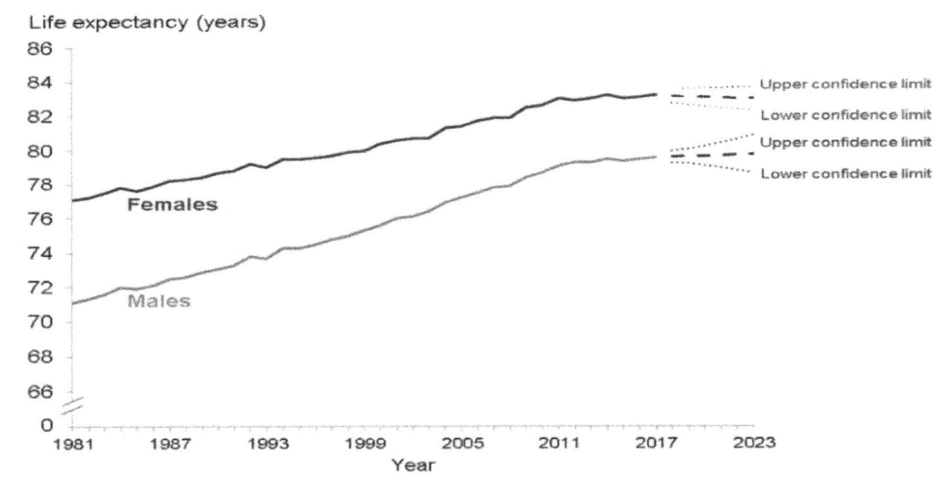

Change and trends in life expectancy[179]

---

[178] Image from BBC web site - Learning English - 6 Minute English *http://www.bbc.co.uk/learningenglish/thai/features/6-minute-english/ep-170302*

[179] Public Health England analysis of ONS data *https://www.gov.uk/government/publications/health-profile-for-england-2018/chapter-1-population-change-and-trends-in-life-expectancy*

Importantly, although we know, deep down, what's coming, we have time to adapt to the changes gradually, as they draw near, trying to stave them off a little, by dying grey patches in our hair, exercising vigorously, covering wrinkles with exotic potions or, perhaps, indulging in a little Botox or even cosmetic surgery. Our society rails against ageing and we don't handle it, or death, well despite the inevitability of both.

The 'Health state life expectancies, UK: 2014 to 2016' [180](ONS) usefully makes the distinction between expectancy of a healthy and an unhealthy life.

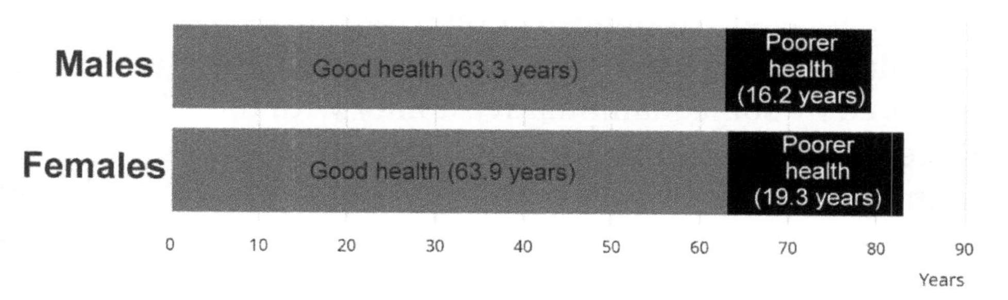

I suspect that most people think only about the absolute life expectancy of >80 years, when they strive to add a few extra years, and not about the final 16 or 19 years of (statistically) poor health. Personally, I'd welcome another few years as an 18 year-old, but I'm in no rush to add time in my dotage. As George Bernard Shaw said ***'youth is wasted on the young'***.

One sign that there had been some catastrophic event in my brain was when I sat down at my PC back at home to perform my habitual check of emails – albeit after a couple

[180] Based on a survey where respondents who answered their general health as 'very good' and 'good' were classified as having 'Good' health. Those who answered 'fair' 'bad' and 'very bad' were classified as having 'Not good' health. https://www.gov.uk/government/publications/health-profile-for-england-2018/chapter-1-population-change-and-trends-in-life-expectancy

of weeks' gap. I couldn't use my right hand, of course, but I was surprised to discover that I simply couldn't remember my password. It took a day or two to remember it and then to go through the "routine", and I worried how much else I might have lost.

After a couple of weeks of fruitless staring at my right hand and trying to will life into it, I noticed that when I yawned my right arm lifted slightly, suggesting that there was some kind of control link. I practiced the stimulus over and over, until I could replicate it at will and felt at least some degree of kinship with the arm again. Encouraged by that success I then managed (though I don't recall using the same technique) over weeks to make individual fingers move, followed by some manipulative ability with the whole hand. Once the connections had been reformed it was easier to improve, but never fully. These were heady days of slow but real progress. It helped that shortly after my return home the Wimbledon tennis championships were on the TV continuously for a fortnight. Not since I began work had I been able to watch it all, and I played every shot in my mind. Although, because of my limited use of right arm and hand, and very limited use of my leg, I couldn't actually play tennis, I reckoned that playing mentally ought to at least stimulate the memory and brain-limb connections.

But it wasn't imaginary tennis that caused me the extreme fatigue which is a well-known symptom of stroke. I would begin each new day at my best, and my 'battery', as I would call it, would progressively run down until I became dyspraxic and my speech incoherent – usually by early evening. After a while I learned which activities placed the biggest drain on my 'battery'. Social interactions in new groups or with particular individuals ranked highest, with any stressful activity or one calling for intense concentration and physical activity or dexterity also very draining. I discovered too that even a very small quantity of alcohol had a similar effect, so I have had to become almost teetotal.

The only way that I could recover from 'battery fade' was a good, full-night's sleep. An afternoon nap won't cut it. The dreams that I remembered were weird and novel, involving endless searching through teak-panelled 1960's office buildings over weeks and months. There were miles of corridors, a couple of lifts which allowed access to only certain floors, and a very long escalator leading from a large, open lobby area (I don't remember ever travelling up or down it, just gazing at it). There were no other people in the building, nor windows, and I have no idea what I was searching for (I never found it). Later, I assumed that my brain had been discovering and mapping damaged connections. Who knows? In any case, in all my dreams I still had all of my pre-stroke mobility; I could walk, run and jump as well as ever. Although that made those dreams tremendously enjoyable, it was a painful shock when I awoke every morning to discover the reality - that I still couldn't leap out of bed, trot downstairs and then off for a game of tennis. That bitter-sweet experience lasted for about 18 months.

I was fortunate to have, in addition to a few weeks' help from the local NHS stroke physiotherapy team, a new toddling Granddaughter making early forays into the exciting new world and simultaneously a Mother-in-law's retreat from it. I was introduced to the thrills of 'Rhymetime' and joined in enthusiastically with the songs and activities from the sidelines. It was the first time since the stroke that I had tried jumping and clapping and was surprised to discover that I could do neither.

But, as the months passed, progress plateaued. I never, for example, regained sufficient fine control of my right arm and hand to be able to write or type, nor to shake drips off my right hand after washing, to kick a football, to operate hand tools, such as a hammer or a saw, speak coherently or to draw or paint. Which is a shame, because I used to enjoy and pride myself on my carpentry, handwriting and drawing skills. Oddly, I seem to fare better with double-handed activities, so whilst I can't operate a handsaw, I can handle a chainsaw. Crucially, from a practical perspective, I cannot hold a walking stick usefully in my right hand to aid walking, but I can drive. What little progress I was making was generally the result of work-arounds rather than regaining previous skills or abilities. Other than using my other hand to wipe my bum, I had to develop new techniques for almost all everyday activities, like dressing, getting in and out of the bath or shower, and controlling cutlery. The landlady of my local pub went as far as to change all the cutlery because I found it difficult to manage! Old friends - who knew the 'old me' - were especially treasured, but almost everyone was kind, understanding and helpful. I particularly remember an evening when I went to meet a couple of old friends in a pub. It was a fairly new outing for me; an adventure. When I arrived at the pub I was confronted by steps, and realised that I was about as agile as a Dalek. Two chaps who were having cigarettes outside the entrance doors very quickly and kindly grabbed one elbow each and hoisted me up and in. With enormous gratitude and a damp eye I realised that I had officially joined the ranks of the elderly, infirm and disabled. In the blink of an eye I had gone from being an instinctive helper of mums with babies in pushchairs on London's underground to a frail old man who needed help <u>into</u> a pub.

The 'Burrell Arms' steps, which defeated me, but for the kindness of strangers

Echoing in my mind, however, was some of the best advice I had had since the stroke – this from one of my physiotherapists – ***'it's easy to fall into the trap of victimhood. Try to avoid it'.*** I interpreted it as a plea to persist with a default view that you can do something until proven otherwise. Sadly it has often proved otherwise, but it's still one of the best pieces of advice I've been given.

There is plenty of material on YouTube about strokes, including some from medical professionals who gained direct personal experience through suffering a stroke themselves. In 1996 Jill Bolte Taylor, a neuroanatomist and American postdoctoral fellow at Harvard Medical School, suffered a major stroke herself, so was able to experience and, more importantly, understand and later describe the effects. In March 2008 she gave a moving and entertaining TEDx[181] Talk[182] - 'My stroke of insight' (also the title of her book, later that same year) - about how she watched as her brain functions - motion, speech, self-awareness - shut down one by one. In 2013 Aphasiologist[183] Dr. Robert Goldfarb also suffered a stroke and gave an entertaining and informative TEDx Talk[184] - 'An Aphasiologist Has a Stroke' - about the experience.

After the allotted few weeks of NHS physiotherapy had finished and I and other survivors had had a while to fend for ourselves, a 'moving on group' was formed and half a dozen meetings held. It was an opportunity, after relative isolation, to compare notes with others who shared experiences which I and several others of the ten or a dozen attendees found particularly helpful. It was refreshing not having to keep explaining and to be able to share useful tips for anyone who was struggling to come to terms with their lot. I have since, at the local NHS OT's[185] invitation attended all subsequent groups of new stroke survivors in a support capacity. It feels helpful, but can be tough – playing along with the optimistic line that recovery will come, and helping individuals through the inevitable process of mourning which the stroke-induced loss of faculties represents. Swiss-American psychiatrist Elisabeth Kübler Ross proposed the "Five Stages of Grief" model based on her work with terminally ill patients and,

---

[181] TED Conferences is an American-Canadian non-profit media organization that posts international talks online for free distribution. To date, more than 13,000 TEDx events have been held in at least 150 countries https://www.ted.com/

[182] 'My stroke of insight' https://youtu.be/UyyjU8fzEYU

[183] Aphasiology is the study of language impairment usually resulting from brain damage

[184] 'An Aphasiologist Has a Stroke' https://youtu.be/LLhXxBC9xYk

[185] Occupational Therapist

although the cause might be different to that on which she worked, the sense of grief is palpable

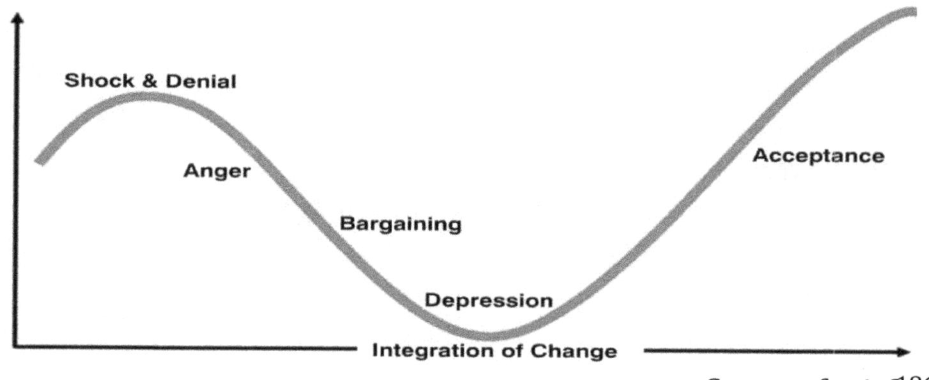

Stages of grief[186]

Even though the stages aren't necessarily all "essential" nor sequential it's useful shorthand and helps identify where each individual is on the path and what their current key challenges are. I've seen many people still suffering from shock and anger or denial and I know that they won't be able to 'move on' until they are through that. But in their own time; with empathetic help, rather than a grab of the shoulders and a vigorous shake. And who knows - they might even beat the odds and make a full recovery?

For me, as a previously keen and active DIYer, sportsman and artist, there was also another set of (additional) phases which I went through:

Tractability

Family accusations that I had a 'controlling' nature had stung, but I had always felt capable of meeting any challenge, and the heavy hand of duty on my shoulder.

---

[186] Based on the 5-stage model introduced in her 1969 book 'On death and dying' by Elisabeth Kübler-Ross

Suddenly this had to change. In the short-term I needed to relinquish all control and to rely on others, and I was grateful for and comfortable with that. But it didn't mean that I didn't aim to regain those capabilities, so I sought and followed advice, working hard on rehabilitation with the aid of a couple of baked bean tins, some resistance bands, and guidance from physiotherapists. Everything else took a back seat. After about six months progress had slowed to a virtual halt and it became clear that a change in tactics was needed.

Postponing

Although I needed occasionally to wield a hammer, spanner, drill or saw for urgent household repairs my ineptitude was frustrating for me (is this what it's like for some people at their best to be so ham-fisted?) and comical to casual observers, so most jobs were put off until the recovery that I was confident I would make. It was just taking longer than I had expected. Meanwhile my level of fitness was suffering. A couple of games of 'badminton' with another stroke survivor and two patient sympathisers (unintended pun), a couple of forays up a loft ladder and a bit of mental tennis doesn't stop your muscles from atrophying.

Impatience

As more time passed the list of outstanding jobs piled up and so did the rolls of fat around my waistline. I realised that if there was going to be any real further improvement then I was going to have to find a new way to make it happen. There was no follow-up contact from my neurosurgeon(s), the physiotherapists had done their bit and moved on, and conventional wisdom that almost all recovery would be achieved during the six months following a stroke seemed to be proving accurate.

Because my brain still seemed to work, I was arrogant enough to believe that I could **think** myself better; that I could use the bits I had which were still working, and which had served me pretty well so far. The first step, I thought, was to better understand brains and strokes. As soon as I was able to do so I obtained from the hospital copies of the scans taken, expecting to see a 'smoking gun' black hole, or similar. But the damage was not immediately apparent, and I didn't know where to look, nor what to look for. Despite there being plenty of (rudimentary) 'how to read brain scans' presentations on YouTube, and now being armed with 6,453 pictures of my brain, I still couldn't spot the damage. Apparently Radiology requires proper training.

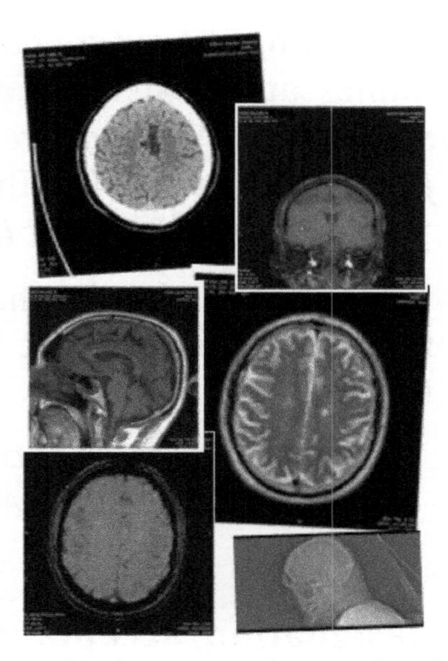

Although I couldn't **see** the damage I am conscious of it, and I know that Small component failure or omission can have catastrophic effects, such as with a hairline crack in a printed circuit board, or a missing rotor arm from a car's distributor.

I spent a lot of time concentrating on every component of every physical activity. Every muscle movement isolated, each step carefully paced. I persisted with it for months and months, voraciously reading a whole new genre of articles and books. For example, in 2014 Yang Wang and Manoj Srinivasan of the Movement Lab at Ohio State University published their findings[187] about the mechanics of walking. Their conclusion was that walking is really just falling and "catching" yourself. Srinivasan says. ***'These series of errors and recoveries are pretty sub-conscious for a healthy adult.'*** But, he says, that to stay safe, maybe ***'you can't think about it too much.'***

I've listed the books that I read (and given authors credit for specific points that I use where I can directly attribute the source).

Nothing is a waste of time if you use the experience wisely (Auguste Rodin)

Resignation/acceptance

(OK – I know Elisabeth Kübler Ross got there before me)

It is now almost six years since I suffered the stroke and I believe that the adage 'experience is what you tend to have just after you needed it' is especially relevant in my case. I suspect that had I known then what I know now I might have made a fuller recovery, or at least wasted less time in futile hope. I don't expect to make any further recovery, and increasingly turn my attention to what I can still do, like writing this book. It's also an opportunity to allow Roger Federer, Ronnie O'Sullivan and Usain Bolt – none of whom ever beat me - to rest easy; I'm officially hanging up my racquet, cue and spikes.

[187] https://royalsocietypublishing.org/doi/10.1098/rsbl.2014.0405

Following are some of the areas that I discovered and found of particular interest, which I hope others might also find interesting. They might even prove helpful to others who suffer a stroke or similar.

Strokes

'Be grateful for the things and people you have in your life. Things you take for granted someone else is praying for'
- Marlan Rico Lee

I've already mentioned strokes briefly under 'I should've expected that' above, but let's give a little more detail here.

A stroke happens when the blood supply to part of the brain is cut off, killing brain cells. There are three different types of stroke:

An ischaemic stroke - caused by a blockage cutting off the blood supply to the brain. This is the most common type of stroke and the one which I suffered.

A haemorrhagic stroke - caused by bleeding in or around the brain.

A transient ischaemic attack or **TIA** - also known as a mini-stroke, it is the same as a stroke, except that the symptoms only last for a short time because the blockage that stops the blood getting to your brain is temporary.

Early treatment is vital to minimise permanent brain damage (hence the FAST[188] acronym), though the Thrombolysis – "clot buster" medicine often needed to treat ischaemic strokes cannot be given before it is diagnosed as such, because it would likely have catastrophic effects for a haemorrhagic stroke sufferer.

---

[188] Developed in the UK in 1998 by a group of stroke physicians, ambulance personnel, and an emergency department physician - **F**acial drooping, **A**rm weakness, **S**peech difficulties and **T**ime to call emergency services

In France (probably elsewhere too) strokes are known as AVCs (Accident Vasculaire Cérébral). A French electrician whom I know suffered an AVC when he fell from a ladder whilst working.

Strokes are common. There are more than 100,000 in the UK each year - one every five minutes. More than 400 children have a stroke every year in the UK. Strokes are the fourth biggest killer in the in the UK. In 2016, almost 38,000 people died of stroke in the UK. That's a life lost every 13 minutes. 1 in 14 deaths in the UK are caused by stroke. 1 in 8 is fatal within the first 30 days. In 2016 strokes caused almost twice as many deaths in women as breast cancer and 5,000 more deaths a year than prostate cancer in men. The effects of a stroke depend on where it takes place in the brain and how big the damaged area is. Stroke is a leading cause of disability in the UK. Almost two thirds of stroke survivors in England, Wales and Northern Ireland leave hospital with a disability. There are over 1.3 million stroke survivors in the UK.

Worldwide someone will suffer from a stroke every two seconds. In 2016, there were almost 14 million incidences of first-time strokes worldwide. They cause around 6.2 million deaths each year, taking a life every five seconds. Almost 1 in 8 (12%) of deaths worldwide are caused by stroke and stroke-related illness, disability and early death is set to double by 2035.

Some of the effects of stroke are:

- weakness in arms and legs
- problems with speaking, understanding, reading and writing
- swallowing problems
- vision problems

- losing bowel and bladder control
- pain and headaches
- fatigue – tiredness that does not go away with rest
- problems with memory and thinking
- numb skin, pins and needles.

Stroke is the second leading cause of death worldwide but I admit that before I suffered one I scarcely gave the risk a thought. It certainly wasn't on my risk list, and those that were focused predominantly on killers, rather than disablers. Like the majority of people, I suspect, I ought to have been more aware of the **healthy** life expectancy data mentioned previously.

The human brain

'Man is the lowest-cost, 150-pound, nonlinear, all-purpose computing system which can be mass-produced by unskilled labor' - S.Fred Singer[189]

Let's begin with a few basic facts about the human brain.

The human brain is the product of about 3.7 billion years of development of which the most recent 300,000 years has been in the skulls of Homo sapiens, or humans. That's about 12,000 generations of Darwinian evolution of the human brain. It's now the most intelligent thing we know of in the Universe, and it is the way it is because it works.

The Scientific American tells us[190] that 'the average male has a brain volume of 1,274cm^3 and that the average female brain measured 1,131cm^3, both reaching their peak size in our twenties, and that 'bigger is slightly better'. That sounds like a recipe for a specious argument that I'd rather avoid, so I'll just say that it typically weighs about 3Lbs, or 1.5Kg - about 2 percent of a human's body weight. It receives information about its environment via a series of sensory organs, which it needs to interpret, because it lives inside a dark cave, called the skull.

It contains about 86 billion nerve cells (neurons) - 'grey matter', billions of nerve fibres (axons and dendrites) – 'white matter' and trillions of connections, or synapses between neurons. In a single cubic millimetre of brain tissue there are one hundred million synaptic connections

[189] in a 1965 article "The Case for Man in Space" in a magazine called "The Reporter"

[190] https://www.scientificamerican.com/article/does-brain-size-matter1/#:~:text=An%20MRI%20study%20of%2046,any%20of%20it%20spilling%20out.

and in a cubic centimetre there are as many connections as there are stars in the Milky Way galaxy. In computing terms that is technically known as 'a lot of grunt'. In May 1997, IBM's 'Deep Blue[191]' finally became the first computer to beat a reigning world champion at chess, when it beat Gary Kasparov 3½–2½ at the second attempt (following an earlier 4-2 loss and an upgrade). Kilowatts of computing power ranged against Kasparov's 20Watts in a 64-square 2-D board game.

The Web is replete with diagrams and presentations about the architecture of the brain. This, from The Khan Academy (who present plenty) is an excellent example, showing principal functions mapped to physical areas.

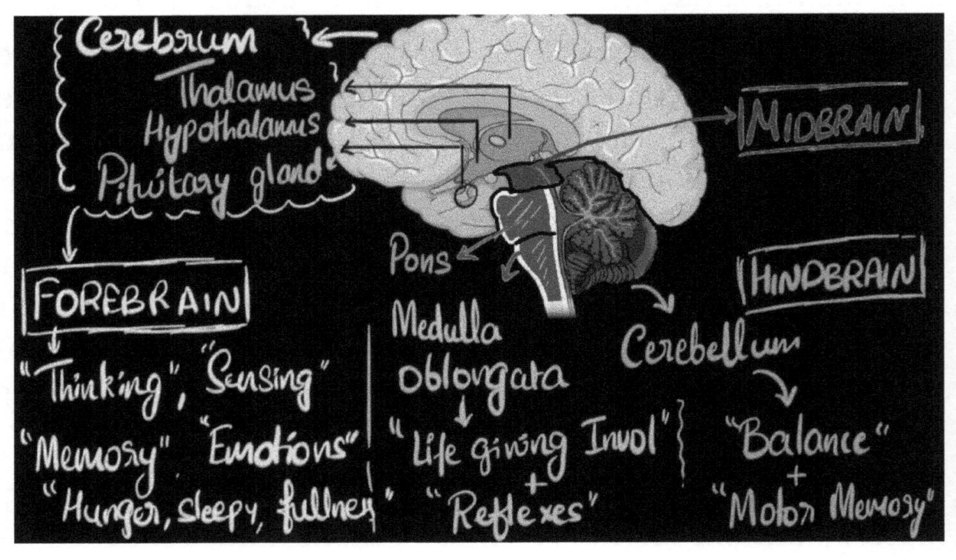

Brain: Parts & functions – The Khan Academy [192]

[191] IBM were known, in the trade, as 'Big Blue'

[192] Adapted from this presentation https://youtu.be/DtkRGbTp1s8

There seems to be broad agreement about where in the brain the various basic functions are performed – presumably supported by increasing evidence from MRI and other clever scanning technology – but there's still a lot to learn, and much still seems a bit woolly to me.

All bilaterally symmetrical animals, like humans, have the same basic brain configuration of Forebrain, Midbrain, and Hindbrain. The Hindbrain looks after all the essential "involuntary" functions, such as breathing and swallowing. The human brain has two hemispheres – left and right - that appear (and mostly are) identical, and which control and receive input from the opposite side of the body (technically called 'contralateral organisation'). So the left side of your brain controls the right side of your body and vice-versa. Some functions, such as speech, are served by specific sides of the brain, and you'll see talk (though little evidence) about potential 'neuroplasticity' enabling rewiring of those functions from damaged to undamaged areas. The two halves of the brain are connected by a thick bundle of nerve fibres called the Corpus Callosum that allow them to communicate with one-another. You'd be forgiven for thinking that this is to provide back-up to cope with damage to one side, but this doesn't seem to be the case. And rather incredibly you seem to be able to survive pretty well with only one half. In order to resolve a rare intractable form of epilepsy, an operation – called a hemispherectomy – to remove one half of the brain can be performed. So long as it's done before the patient is about eight then they will be fine, and you probably wouldn't be able to guess. It's led me to wonder about some of the people I've met over the years. Anyway, I'm sure I don't need to tell you not to try this at home, unless of course, you happen to be a qualified neurosurgeon, and do make sure you have the patient's permission first.

David Eagleman argues – persuasively, I think - in his book 'Incognito' for a 'team of rivals', where the brain contains various specialised modules which can argue with itself; laugh at itself; and generally be in conflict with itself; torn on single issues. It fits well with the need for differences from which Darwinian improvement can emerge, and the seemingly strange twin brain hemispheres.

Immediately underneath the Corpus Callosum is an area called the Basal Ganglia which is even more mysterious, but seems to play an important role in all sorts of things including motor control (and so is of key interest in the case of Parkinson's disease). Together with the Cerebellum – the thing which looks a bit like a "mini brain" and where learned motor action, like riding a bike or serving a tennis ball, are embedded – it is also of particular interest to me. I can't fathom why suddenly I can't perform a host of familiar actions, like walking, writing and drawing or buttering a slice of toast. Oh, and it's the Cerebellum which is affected by alcohol, apparently. But it doesn't explain my problems with speech, which conventional wisdom says is controlled by an area low down on the left hemisphere of the Forebrain called Broca's area, in the frontal cortex above and behind the left eye. That certainly fits with the pains I suffered in the lead-up to the stroke.

Steven Pinker offered some useful insights in his book 'How the mind works' (1997):

> ***'The various problems for our ancestors were subtasks of one big problem for their genes, maximizing the number of copies that made it into the next generation.'***

He then goes on to describe the modularity which must represent a key design feature of the human brain in pursuit of those overall goals:

> ***'The mind has to be built out of specialized parts because it has to solve specialized problems.'***
>
> ***'Whether or not we establish exact boundaries for the components of the mind, it is clear that it is not made of mental Spam but has a heterogeneous structure of many specialized parts.'***

Which I think means that neuroplasticity is confined to specialised parts, and that you can't just use a bit of spare capacity in one to replace something broken in another.

In my humble opinion although the "brain functionality maps" are interesting and helpful they only tell part of the story. Just as with Function Models of an Enterprise, understanding the interactions between discrete functional "modules" is as important. A key 'specialised part' is memory and I know, for example, that somewhere within the undamaged parts of my brain there is crystal clear memory of walking, speaking, writing, drawing and so on. I can feel, hear, smell, and experience them, both consciously and whilst asleep, dreaming. And I still have complete control over half of my body (albeit the opposite side to my natural handed side. I could never write or draw with my left hand, nor kick a ball with my left foot). If there is a "back-up", then why can't I just switch to it? It has to be something else – something more subtle.

Memory

'Life is all memory, except for the one present moment that goes by you so quickly you hardly catch it going' - Tennessee Williams

Memory is an area where there is a stark divergence between humans and computers. True, computers have short-term, working memory (Random Access Memory - RAM) and longer-term memory in the form of disk or Solid State Memory (SSD). But the substance of what is stored, and the form in which it is stored is entirely different. Instead of bits and bytes of binary data our memory stores the rich tapestry of senses in which the events were originally experienced. The sounds, smells, feel, emotions and other sensations such as temperature, motion, and the context in which it happened. When discussing language – the medium we use to share experiences - Stephen Pinker calls it 'Mentalese', which he describes as

> ***'The hypothetical language of thought, or representation of concepts and propositions in the brain, in which ideas, including the meanings of words and sentences, are couched.'***

The fundamental premise is, I think, that in order to communicate about something you need first to be able to understand and articulate it internally – to have it clear in your own mind, including not just current and past experience but also potential future experience, probabilities, options and much more. "What DID I just see", and "will going next Tuesday to the same spot on the beach as we did last month be as much fun?" for examples. When it comes to expressing a memory to others it might require a bit of disambiguation, beyond grammatical clarity – "I saw the girl with binoculars on the hillside", for example. And think about the ways in which language tries to convey experiences or concepts. Take as an example the transfer water onto a surface. Splash, squirt, spray, drip, pour, splatter, sprinkle are just a few of the words that we have available to help us to communicate about a concept which we've probably experienced. As you read each you'll probably have an image in your mind – maybe also accompanied by a sound.

I couldn't hope to do justice to the subject, and I would urge anyone who is interested in the complex and fascinating subject of language to read Stephen Pinker's books.

Returning to the question of short-/long-term memory, most people don't understand what 'short-term memory' is. The human brain can hold only four 'items' at any one time and only for up to about a minute but, as Dean Burnett explains, in his book 'The idiot brain: a neuroscientist explains what your head is really up to (2016)', People have

> ***'developed strategies to get around limited short-term-memory capacity and maximise available storage space. One of these is a process called 'chunking', where a person groups things together into a single item, or 'chunk', to better utilise their short-term memory capacity. If you were asked to remember the words 'smells', 'mum', 'cheese', 'of', and 'your', that would be five items. However, if you were asked to remember the phrase 'Your mum smells of cheese', that would be one item, and a possible fight with the experimenter.'***

The purpose of short-term memory is to process, in real-time, information coming in from sensory organs and long-term memory incredibly quickly. It's the multiple inputs which can sometimes give rise to confusion, especially in the case of dementia – what is "real" and what is not. And it's easy to see how we can, with competitive vying for our attention, sometimes find ourselves standing in the kitchen wondering "now, why did I come in here?" It's also, I believe, the seat of our 'mind's eye'.

Long-term memory is, by contrast, in Burnett's words ***'obscenely capacious' 'nobody has lived long enough to fill it'*** and can best be thought of as a reference library.

Memory encoding is done by the Hippocampus, regulated by the amygdala – the brain's 'emotional memory processor', before being stored in the frontal lobe of the brain (where, conventional wisdom tells us, most 'thinking' is done). The amygdala is critically involved in enabling us to acquire and retain lasting memories - making "decisions" about preservation worthiness of emotional experiences in particular, or where a shift in perception is required.

The Hippocampus (so named by the Greeks after a seahorse, which it resembles) also has an important role in spatial awareness and manages retrieval of experiences from the frontal lobe. Interestingly, London taxi drivers (who have 'the knowledge') have been found to have significantly larger Hippocampi than control subjects, suggesting either that people with a large Hippocampus are drawn to taxi driving or, more likely, a capacity for neuroplasticity in that region of the brain.

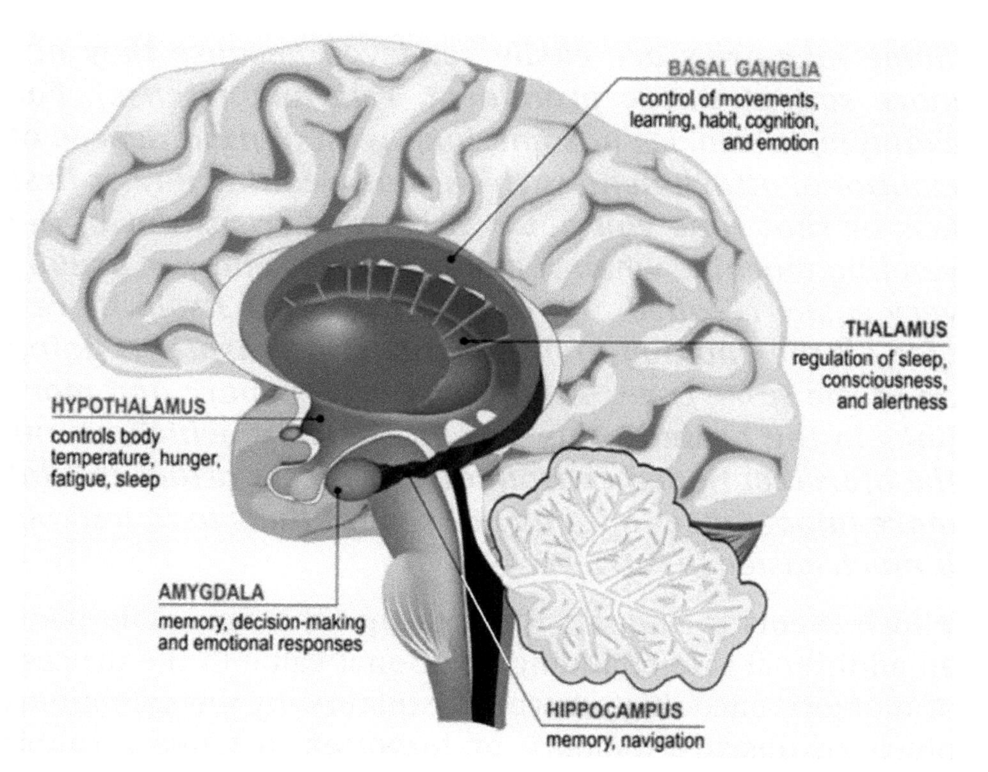

Limbic system - (Illustration credit: Designua / Shutterstock)[193]

Additionally, as Dean Burnett explains:

New memories are laid down by the hippocampus and slowly move out into the cortex as new memories form 'behind' them, gradually nudging them along. This gradual reinforcing and shoring up of encoded memories is known as 'consolidation'. 'So, if the long-term memory remembers everything, how do we still end up forgetting things? Good question. The general consensus is that forgotten long-term memories are still technically there in the brain, barring some trauma in which they're physically destroyed. But if you can't retrieve a memory, it's as good as not being there at all.'

[193] via News-Medical - https://www.news-medical.net/health/Hippocampus-Functions.aspx#3

'Some memories are easily retrieved because they are more salient (more prominent, relevant, intense). For example, memories of something with a great degree of emotional attachment, such as your wedding day or first kiss or that time you got two bags of crisps out of the vending machine when you only paid for one, are usually very easily recalled. As well as the event itself, there's also all the emotion and thoughts and sensations going on at the same time. All of these create more and more links in the brain to this specific memory, which means the aforementioned consolidation process attaches a lot more importance to it and adds more links to it, making it much easier to retrieve.

Which seems to support the idea that the Amygdala plays an additional role of adding emotional value in the process of memory encoding. It would explain why we sometimes pluck an obscure memory or forgotten fact like a rabbit from a magician's hat. Consolidating memory is one of the brain's key activities while we sleep – about which more later.

I've already mentioned the grammar inherent in language, which is a key structural component for meaning, tenses and concepts, and the brain likely uses a number of other, simpler devices to help with encoding experiences. As well as linguistic grammar we learn, from a very early age, other "grammars", such as of colour (palettes), of motion sensations, of mathematics, of smells, of tastes, of sounds (beat, melody, instruments, volume etc.), of touch/texture (rough, smooth, soft, sharp...), of temperature, of form (Euclidian geometry, geons. Geons are simple 2D or 3D forms such as cylinders, bricks, wedges, cones, circles and rectangles corresponding to the simple parts of an object.

There are fewer than 40, from which all complex items can be constructed. Irving Biederman developed a 'Recognition by Components Theory' in which he proposed that the visual input is matched against structural representations of objects in the brain. It certainly applies for my dog, Bertie, to whom a spherical paperweight, stone, or an orange, are all balls, which should immediately be thrown for him to chase).

Pinker says that the human brain is primed to learn language by the age of six (which helps to explain why I struggle to learn French now) and it's encouraging to hear a toddler announce that they have 'breaked' something, demonstrating that they have grasped English grammar, if not vocabulary. That's quite an impressive feat, and I wonder how many of the other 'grammars' mentioned above they have also mastered by that age?

I have no idea what the process for encoding the special "expertise" into the Cerebellum is, but I make no apology, since I've barely begun my neuroscience study.

Seeing is believing

'Sight is the noblest sense of man' - *Albrecht Durer*

'Art is a line around your thoughts' - *Gustav Klimt*

'The only thing worse than being blind is having sight but no vision' - *Helen Keller*

Human eyes are incredible feats of evolution, but they aren't a patch on those of raptors like Hawks, and certainly wouldn't pass muster as precision optical instruments. As the nineteenth-century German scientist Hermann von Helmholtz said ***'The eye has every possible defect that can be found in an optical instrument, and even some which are peculiar to itself'.***

Although the range that our eyes can "see" is very wide, most is just a peripheral, blurry view, which gives us awareness of context and of threats. Only at the Fovea – at the centre of the Retina – is our vision crystal clear. If that wasn't the case your eyes wouldn't be skipping along the lines as you read this; you'd simply gaze at the whole page and take it all in without needing to move your eyes. When you stand on a London Underground station platform waiting for the arrival of a train and gazing at the adverts plastered to the opposite tunnel wall your attention will occasionally be caught by a mouse scurrying in your peripheral vision. You can't <u>see</u> the mouse until you turn your head and direct your fovea at it, but your brain has clocked that there's something unexpected over there.

When it comes to sight it's the brain which does all of the heavy lifting. About a half of its considerable power is employed on seeing, and it allows us to "see" things even if they are not necessarily there.

A prime example is the blind spot which we all have in each eye at the point where the 'wires' from the retina, at the back of the eye, converge into the optic nerve. You don't notice it because the brain fills in the gap in your vision, but you can test the effect very easily. Cover your left eye and keep looking at the star, moving the image closer to you until – voila - the spot magically disappears.

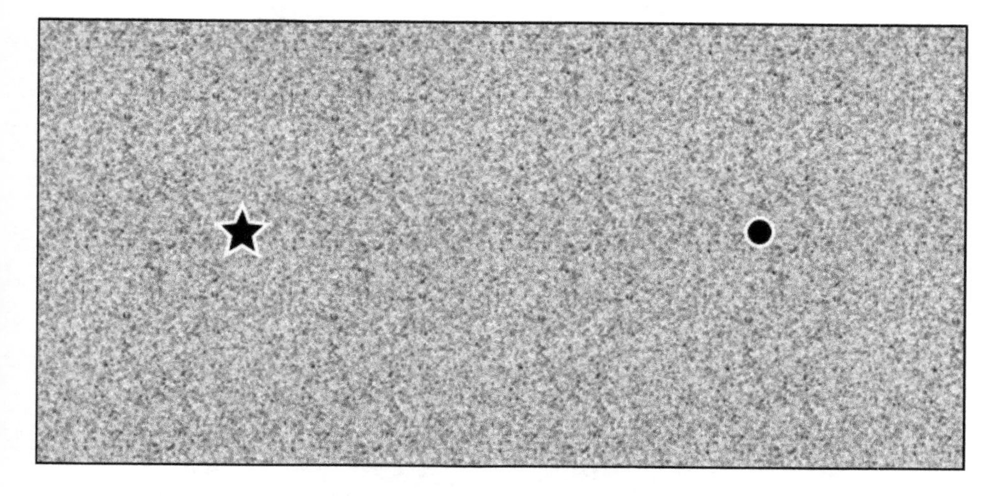

What happens is that the brain simply fills in the missing data. It's just one of a number of tricks and techniques that it employs to make sense of a complex world from the limited visual data fed to it in its dark cave. It does a remarkable job of using existing knowledge and experience of the real world, making assumptions and educated guesses, to translate the simple 2-D images it receives from the eyes into a virtual 3-D world. In doing so it can create objects which, as shown by the following examples, are not really there (top) and make flat images appear to have solid form (bottom)

But it can make "mistakes" and this has provided fertile ground to mess with your mind for decades. You've probably seen some of these 'brain-teasers' before

Dutch graphic artist MC Escher turned it into an art form

And more recently Stereograms or 'Magic Eye' pictures were all the rage for a spell

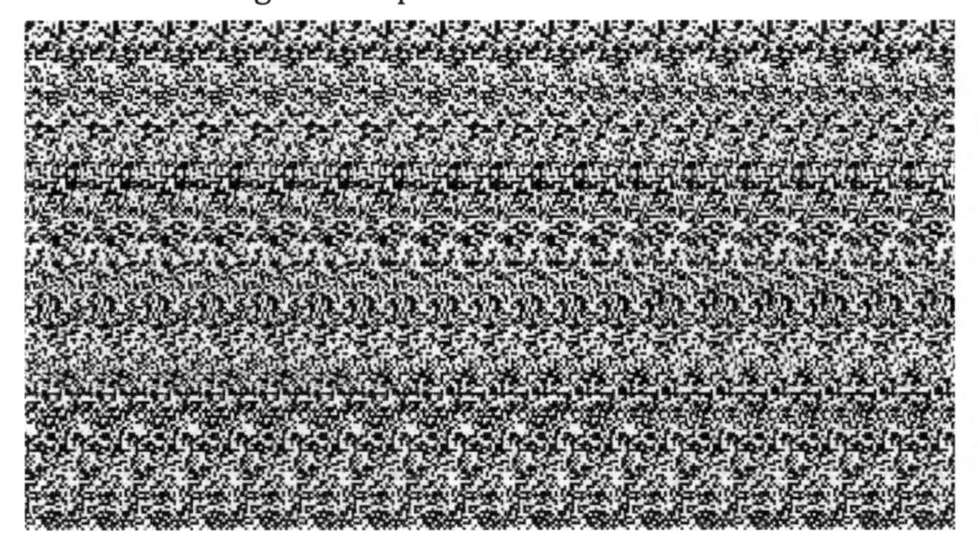

As a keen amateur artist myself I was always curious about how the conventional approach to perspective could be reconciled with what must be a concave mirror view that our eyes give us. How could our 130° field of vision be represented by straight lines? Imagine standing in the middle of a straight railway track (do not actually do this – especially not on any train track in service). Imagine looking down at your feet, where the rails will be parallel. Then look up at the horizon – where 'traditional' perspective theory tells you there will be straight lines converging on a 'vanishing point'. So there must be a curve in the lines somewhere, right? Of course there is the fact that we are only able to focus on a small central area, but even that must have some concave "distortion".

Monet famously accommodates the wide view in his 1920s paintings of his garden and pond at Giverny with very wide (up to 17m) canvasses, meaning that you need to stand in front of them and swivel your head and/or eyes in order to take in the whole vista.

The Water Lilies - The Two Willows (1914 to 1926) 2m x 17m (55'9"), Musée de l'Orangerie

Since I didn't share Monet's skill (and I didn't have access to a 55 foot canvass anyway) my own attempts, over 50 years ago, to interpret the true curved nature of perspective were meagre by comparison. Here's a sketch I did in 1972 where you can at least see the curve of the ceiling joists (which were actually straight).

'A' level art studio sketch, Bill Swan, 1972

I can't offer you a more recent example, because I'm no longer able to control my sketching hand, but Paul Heaston, an artist in Denver, works as an 'urban sketcher' of his surroundings in everyday situations, such as his home or cafés, and he uses the same curved perspective technique, but more skilfully.

Speech

'It is better to remain silent at the risk of being thought a fool, than to talk and remove all doubt of it' - Maurice Switzer[194]

There has been a long line of well-known stars who have suffered from the condition called 'aphasia', including Glen Campbell, Kirk Douglas, Terry Jones and Sharon Stone. Mary Davis Travis, wife of country and western singer Randy Travis – another sufferer – told the LA Times[195] that 'between a third and 40% of the 800,000 people who suffer a stroke in the USA every year is left with the aphasia'.

So what is aphasia? According to the Stroke Association 'Aphasia is a language and communication disorder affecting more than 350,000 people in the UK. It can affect a person's ability to understand speech, speak, read, write and use numbers.' Wikipedia tells us that 'Aphasia is an inability to comprehend or formulate language because of damage to specific brain regions' and that 'the major causes are stroke and head trauma, brain tumours, brain infections, or neurodegenerative diseases (such as dementias)'.

Aphasia is usually caused by damage to specific areas of the left side of the brain, and falls under the following broad categories (see National Aphasia Association[196] for details)

---

[194] often incorrectly attributed to Abraham Lincoln or Mark Twain
[195] https://www.latimes.com/entertainment-arts/story/2022-03-30/bruce-willis-aphasia-emilia-clarke-sharon-stone
[196] www.aphasia.org

Arcuate fasciculus

Broca's Area

Located in **inferior frontal gyrus (Area 44).**

Includes 2 motor areas controlling "Spoken & Written speech".

Function - **Center for motor part of speech & sentence formulation**

Connections - Information received from Wernicke's area → Send to motor cortex.

Lesions →**"Motor/Non-fluent aphasia" - Inability to express oneself by spoken speech.**

Broca's Area

Wernicke's

Wernicke's area

Located in **posterior end of superior temporal gyrus.**

Concerned with COMPREHENSION, planning of words & Integration site for 2° somatic, auditory & visual areas.

Connections - Ultimate signals **send to Broca's area via arcuate fasciculus.**

The speech areas of the brain[197]

Broca's aphasia ('non-fluent aphasia')

In this form of aphasia, speech output is severely reduced and is limited mainly to short utterances of less than four words. Vocabulary access is limited and the formation of sounds is often laborious and clumsy. The person may understand speech relatively well and be able to read, but be limited in writing. Broca's aphasia is often referred to as a 'non fluent aphasia' because of the halting and effortful quality of speech[198].

Wernicke's aphasia ('fluent aphasia')

In this form of aphasia the ability to grasp the meaning of spoken words is chiefly impaired, while the ease of producing connected speech is not much affected. Therefore Wernicke's aphasia is referred to as a 'fluent aphasia.' However, speech is far from normal. Sentences do not hang together and irrelevant words intrude-

[197] Taken from Medicoapps excellent explainer. You can see it – with examples – at https://youtu.be/G9SZJxjDJpc

[198] https://www.neuroscientificallychallenged.com/blog/know-your-brain-brocas-area

sometimes to the point of jargon, in severe cases. Reading and writing are often severely impaired[199].

Despite what you might be thinking as you read this, the aphasia which I suffer from doesn't fit these definitions neatly. I don't suffer from Wernicke's aphasia, and my Broca's aphasia is relatively mild. Until I become tired, or stressed, or consume even a very modest amount of alcohol. Then I can have extreme difficulty forming words, and I become so incoherent that even I don't know what I'm talking about. My Wife points out that that was often the case before I suffered a stroke.

Because I can think speech perfectly well - comprehension, grammar and so on – it is particularly frustrating when I can't get the words out. The same is true for writing and drawing. I used to be proud of my writing, but cannot now make my hand direct the pen or pencil where I want it to go. It's weird, and especially frustrating. I suspect it's more due to loss of fine-motor skills than damage to specific speech-related areas of the brain.

---

[199] https://www.neuroscientificallychallenged.com/blog/know-your-brain-wernickes-area

Autopilot

'What a piece of work is man! How noble in reason! How infinite in faculties! In form and moving, how express and admirable! In action how like an angel! In apprehension, how like a god! The beauty of the world! The paragon of animals! And yet, to me, what is this quintessence of dust?' – Hamlet, William Shakespeare

I used to be a keen amateur tennis player. I loved it and I miss it. I remember going to watch the championships at Wimbledon in the late 1970s and, in particular, watching Roscoe Tanner from a corner vantage point on an outside court. His serve was remarkable – clocked at 153mph in an age of wooden racquets and before jumping serves were allowed. It was almost unplayable but, if the receiver did manage to return, then Tanner became a tennis mortal. And, although he peaked at a world ranking of 4 (in 1979), won the Australian men's singles title in 1977, reached the Wimbledon men's singles final in 1979 (where he lost to Bjorn Borg) they did. How? At 150mph it takes just 0.47 seconds[200] for the tennis ball to travel from the server's racquet to the baseline at the other end of the court, 78 feet away. During that time you need to see when the ball is played, the direction and trajectory in which it is travelling, decide on the shot to play, adjust your position accordingly, draw back your racquet and execute your shot, making adjustments for bounce and the server's court position. All in less than half a second. But opponents did it sufficiently frequently to limit Tanner to a single Grand Slam title. That wasn't just the result of serving faults.

[200] https://tennisone.tennisplayer.net/club/lessons/brody/serve/150.php

I didn't have such an acute interest at the time, but I sensed that when he wrote his article 'The Art of Failure' for The New Yorker in August 2000 Malcolm Gladwell was on to something. The article was about 'choking' in sport, and used sporting examples such as that of Jana Novotna in the 1993 Wimbledon Ladies singles final, when she crumpled from a 4-1 final set lead to lose to Steffi Graf.

Gladwell references a study by Daniel Willingham, a psychologist at the University of Virginia, about implicit learning where, after a great deal of practice at something there comes a point where you can do it without thinking. Using the example of playing tennis shots Willingham says that

'when you are first taught something — say, how to hit a backhand or an over-head forehand — you think it through in a very deliberate, mechanical manner. But as you get better, the implicit system takes over: you start to hit a backhand fluidly, without thinking. When that system kicks in, you begin to develop touch and accuracy, the ability to hit a drop shot or place a serve at a hundred miles per hour. This is something that is going to happen gradually. You hit several thousand forehands, after a while you may still be attending to it, but not very much. In the end, you don't really notice what your hand is doing at all".

As Gladwell said about Novotna

'She lost her fluidity, her touch. She double—faulted on her serves and mis-hit her overheads, the shots that demand the greatest sensitivity in force and timing. She seemed like a different person — playing with the slow, cautious deliberation of a beginner — because, in a sense, she was a beginner again; she was relying on a

learning system that she hadn't used to hit serves and overhead forehands and volleys since she was first taught tennis, as a child'

It took Novotna four years to reach another Wimbledon final, which she lost to Martina Hingis, but the following year, 1998, she won, beating Martina Hingis and Venus Williams along the way.

In 2005 Malcolm Gladwell wrote 'Blink: The Power of Thinking Without Thinking', which examines the way in which the split-second decisions we make unconsciously (intuitively) often prove to be better than those where we spend time and employ extreme conscious thought. I take it to be an extension of the same phenomenon.

In his 2011 book 'Incognito' David Eagleman begins, provocatively

'Almost the entirety of what happens in your mental life is not under your conscious control, and the truth is that it's better this way. Consciousness can take all the credit it wants, but it is best left at the sidelines for most of the decision making that cranks along in your brain.'

Eagleman then describes how, through the process of repetition, the brain converts explicit to implicit knowledge which is 'burned - down into the brain's circuitry' where, he says, 'we can no longer access it. Collectively, these instincts form what we think of as human nature'. Eagleman continues

'When [the brain] meddles in details it doesn't understand, the operation runs less effectively. Once you begin deliberating about where your fingers are jumping on the piano keyboard, you can no longer pull off the piece'.

'The specialised, optimised circuitry of instinct confers all the benefits of speed and energy efficiency, but at the cost of being further away from the reach of conscious access'

And don't forget professor Srinivasan's conclusions about walking; ***'pretty sub-conscious for a healthy adult.' 'to stay safe, maybe 'you can't think about it too much.'*** I guess that, at the time, I got too preoccupied with the interesting physics and didn't pay sufficient attention to this pearl of wisdom at the end, despite the evidence of my own experience; a member of the physiotherapy team who visited in the months immediately following my stroke took me on short walks in the lane near my home. The more I thought about it the worse the dyspraxia became. I came closest with a brief shake-out then an impersonation of Baloo the bear from Disney's 'Jungle Book'.

Eagleman's insights helped me to realise too late that there was a common cause to most of my problems; the loss of previous skills, the inability to perform simple, familiar activities like walking, and the extreme fatigue (because I have to revert to explicit skills and knowledge). And just knowing that helps, somehow. The problem is that the expected corollary that, although it might take a few years, one can simply re-learn those things doesn't seem to apply. I remember watching my Grandchildren and thinking that I could do all that toddling, learning to write, draw, run and general dexterity stuff all over again, and that I could be patient for three of four years. But, after six years, I can't. I guess that, as with learning language, there is a pre-programmed schedule, and if you miss your slot, it's gone. Perhaps he was already aware of this when Aristotle said

'Give me a child until he is 7 and I will show you the man.'

I don't mind that the Grandchildren have overtaken me (it's a delight to watch), but helping to show them how to do these things and more is something else I miss.

Sleep

'If sleep doesn't serve an absolutely vital function, it is the greatest mistake evolution ever made'—Allan Rechschaffen, University of Chicago

If you don't take sleep a lot more seriously once you've read Matthew Walker's book 'Why we sleep' then there must be something wrong with you. Or at least there probably will be in a few years' time. I was so taken with it I looked into whether it was possible to be put into an induced coma for a spell to give my brain some time to fix itself. It only took brief research to discover that while (barbiturate) induced comas have a place in medicine, they also carry serious risk of side-effects (estimated 25%[201]), a 6% - 10% chance of not waking up, and little evidence of the benefits that I was hoping for. So, no.

But sleep is, for me, a vital tonic. A good, full night's sleep is the only way in which I can recharge my "battery", and start a new day afresh, with a degree of linguistic and physical fluency.

I've always appreciated the restorative powers of sleep and recognised the wisdom of the adage 'sleep on it', but my appreciation for it has been heightened since the stroke as I pay more attention to what my brain is up to. How many times have you gone to sleep on the horns of a dilemma, or simply mulling over options, only to wake knowing clearly the answer? I certainly have, on many occasions. It seems to be further evidence of what David Eagleman tells us

[201] https://www.rn.com/nursing-news/medically-induced-coma-inducing-state-of-profound-unconsciousness/

about what our brain gets up to while we're not paying attention.

In Walker's book he explains the benefits of a good night's sleep, gives helpful suggestions on how to get it, and explains what is going on.

Walker describes the processes which the brain goes through whilst we are asleep. The model that he shows illustrates the patterns of sleep, based around 90-minute cycles of sleep which go through REM (Rapid Eye Movement) – when our brain activity is not very different to when we're awake – to four levels of NREM (Non-Rapid Eye Movement) – when most of the body's functions are stopped and various stages of housekeeping are undertaken, including the "weeding" of neural connections.

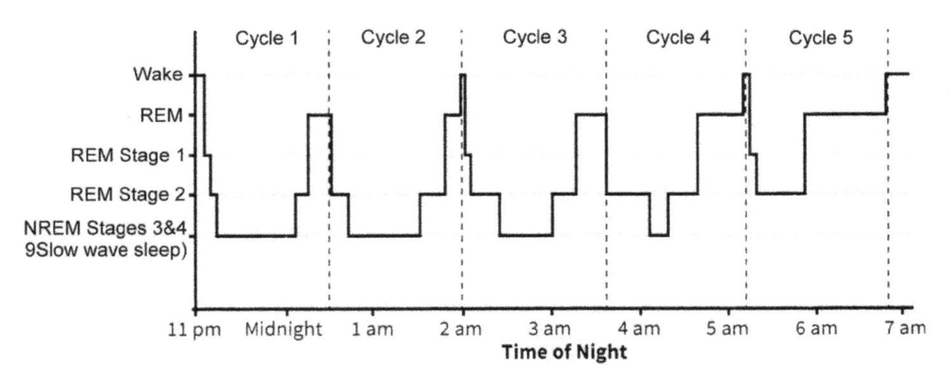

'The Architecture of Sleep'[202]

REM sleep is when we dream and when neural networks are strengthened. You'll notice that this type of sleep predominates later at night.

When you get to my age the three-hourly 'wake' cycle is normally accompanied by a trip to the loo.

[202] From Matthew Walker's book 'Why we Sleep'

The "housekeeping" which the brain undertakes falls into two basic categories, the first of which is the "information management" mentioned above. Computers go through a pale imitation of the same routine – also usually at night, whilst nothing else is going on. Depending on how you have yours set up, it'll clear out unused and temporary files.

Then (if you've asked it to) it'll consolidate files which are left – originally saved as jumbled (but indexed) byte-sized packets of data placed wherever there was space on the disc – into tidier blocks that allow for easier, faster access, and free-up large contiguous blocks for new storage. Along the way it'll make sure there are no broken or fragmented files and, if there are, fix them if it can.

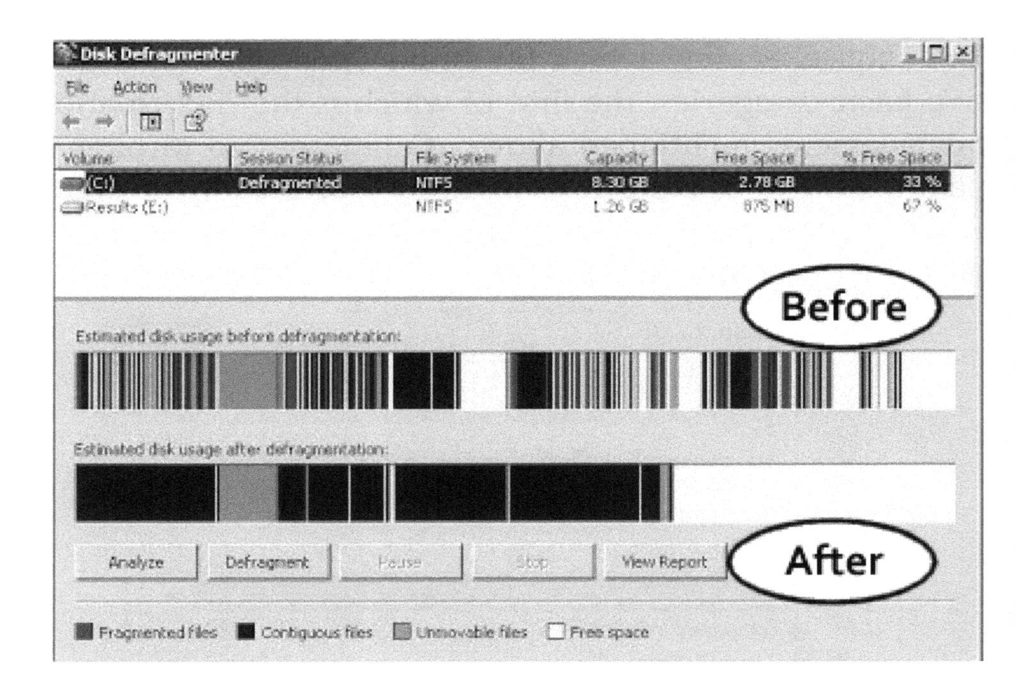

The other kind of "housekeeping" that goes on while you sleep is a thorough physical cleaning. In 2012 Prof Maiken Nedergaard and her colleagues at the University of Rochester Medical Centre, in the US, identified a previously unknown plumbing system in the brain that enables it to clean itself while you sleep[203]. They dubbed it 'the glymphatic system'. During NREM sleep cerebrospinal fluid (CSF) that usually surrounds the brain is pushed around and through the brain tissue in waves, carrying the day's molecular detritus away as it leaves.

Nedergaard believes the system could be important for the clearance of many damaging molecules, including beta-amyloid, a toxic protein that accumulates inside the brains of Alzheimer's patients, and disrupts brain cell function, the tau protein that accumulates in Parkinson's disease, to lactic

[203] From Linda Geddes', Science correspondent report 'The battle to boost our deep sleep – and help stop dementia', The Guardian, 13 Mar 2023

acid, which builds up in the brain when we are awake and which has been linked to seizures, and inflammatory molecules produced by immune cells resident in the brain.

Epilogue

My mother died young. After more than one major operation to remove cancer she suffered several years of pain, discomfort and indignity ending in death at the age of just 64. My father had not long retired, and they were on the brink of enjoying together all the travel and activities which a working and family life had denied them. Then came the diagnosis. Mum believed in the afterlife, where I am sure that she expected to meet her mother and younger sister, who had both already died from the same ailment, and I am sure she wanted to have shared the same experience when they were reunited. So she endured the whole "process" with extraordinary courage.

I don't share her faith - her belief in an afterlife – nor her courage. I like to think I am a pragmatist who can accept when my time is up. They say that it is the hope that dies last, and hope had sustained me through what recovery I had achieved. But they also say that it's the hope that kills you and, after six months, as improvements petered out and it had become clearer what the future held, I was faced with considering what my future could look like, with adjustments (including by others). Since the stroke I had been offered two consultancy engagements, but had already decided that I couldn't do them justice, and declined.

In a parallel universe that spring morning could have been my last. But it wasn't, and I now felt in limbo. The old "me" HAD died that day, and I was in a state of mourning. But a new, different 'me' had survived. Was it enough? Could I live with the frustration of suddenly being able only to do a fraction of what I had been able to do? Would "new me" overwrite, supplant the old me? How would either option affect others? I had been given the opportunity to make a choice. I was 61, with my best days already behind me, declining health to look forward to, and without most of the buoyancy, excitement and energy which new discovery and achievement can give. I had been lucky. I had experienced a lot of things that many others will never have the opportunity to experience. I had visited seven continents, never had to live through a war, nor had to suffer hunger or privation. enjoyed the love of my sweetheart for over 40 years, fathered and played a minor part in the bringing up of two beautiful, intelligent and lovely children, worked for 44 years for 15 employers, in 13 different industries doing things which were useful and/or enjoyable and without knowingly hurting anybody, been blessed with some great friends, had scribblings published in local newspapers and about 9 or 10 books, flown in a helicopter and a hovercraft, driven a Lotus Elise around the Brands Hatch Grand Prix circuit and a Porsche 911 on the road, and achieved relative financial security for my family. I decided this was a good point to stop. Cash my chips. So I set off on what I intended to be my final adventure. I'm here, writing this, because my family found me and brought me home. But that's another story.

While checking dates I realised something interesting from old CVs. The principal purpose of a CV is, of course, to describe an individual's course of life, but it also gives other hints. In mine I had included my 'hobbies and interests' as Carpentry, drawing, tennis and walking. Although I had tried to fit as much of those activities in as possible during free time whilst working, they really represented aspirations for retirement activity.

The lyrics of my favourite band, Pink Floyd's 'Time' resonate all the more:

> ***Tired of lying in the sunshine, staying home to watch the rain.***
> ***You are young and life is long, and there is time to kill today.***
> ***And then one day you find ten years have got behind you.***
> ***No one told you when to run, you missed the starting gun.***

I'm a late convert to the 'treat every day as if it's your last' notion, because one day it will be. I've long held the view that my primary purpose has been to serve as an example to others of how not to do it.

Whilst it's often easy to remember the first time one did something, or a particularly memorable instance (like scoring a great goal, making a hole-in-one, or creating a great painting) it's less easy to remember one's last - especially when it is unexpected. When was my last game of tennis? My last run? Long ramble? Jump? When did I make my final blood donation? Do a tennis ball drop kick for my dog to chase (I'm sure poor Bertie still wonders what awful thing he did on that fateful morning to upset me so much that I haven't done it since)?, Sign my last signature? Paint my final painting? Climb my last tree?

Happier times: A drop-kick for Bertie

If there is a moral here it is to appreciate what you have whilst you have it. Do what you can do, whilst you can do it. However much better you think others are than you at thinking, running, painting, making, playing, singing, you are brilliant. That you can do any of these things is fantastic. If you can do one or more well is wondrous. Celebrate it but, most of all, recognise and enjoy it. Tomorrow it might be gone. And some people might never have experienced it at all.

Remember

- Einstein wrote his theory of Relativity at the age of 26
- Newton was 45 when he published his Principia
- Darwin was still only 50 when the culmination of his ground-breaking work 'On the Origin of Species' was belatedly published
- Matthew Flinders had circumnavigated and mapped Australia by the time he was 28
- Jesus died at 33.
- The best things in life aren't things.

Oh, and get your bucket list done early - don't put it off - do it now, while you can. You might be lucky enough to need to come up with a bucket list #2.

Bibliography and further reading

Where I can I have given specific attributions and credit but, in case I have missed any, following is a list of books which I have been reading during the time that I have been writing this book, or which have influenced my thinking on the subjects covered (in alphabetical order).

A historical look at Enterprise Architecture with John Zachman (The Open Group blog) https://blog.opengroup.org/2015/01/23/a-historical-look-at-enterprise-architecture-with-john-zachman/

'Blink : The Power of Thinking Without Thinking', Malcolm Gladwell, 2005

'Critical mass: how one thing leads to another', Philip Ball, 2004

'Do no harm', Henry Marsh, 2014

'Freakonomics: A Rogue Economist Explores the Hidden Side of Everything', Steven D. Levitt and Stephen J. Dubner, 2007

'How the Mind Works', Stephen Pinker, 1997

'Incognito – the secret lives of the brain', David Eagleman, 2011

'Other Minds: The octopus and the evolution of intelligent life', Peter Godfrey-Smith 2017 (esp. Ch.6: Our minds and others)

'Population change and trends in life expectancy' (Chapter 1), Public Health England Research and analysis, published 11 September 2018 https://www.gov.uk/government/publications/health-profile-for-england-2018/chapter-1-population-change-and-trends-in-life-expectancy

Professor Jack L. Gallant, Berkeley University of California, Psychology gallant@berkeley.edu

'Stepping in the direction of the fall: the next foot placement can be predicted from current upper body state in steady-state walking', Yang Wang and Manoj Srinivasan, 1st September 2014

'The Art of Failure', Malcolm Gladwell, The New Yorker, 21 & 28 August 2000

'The greatest show on Earth'; The evidence for evolution, by Richard Dawkins, 2009

'The idea of the brain: A history', Matthew Cobb, 2020, 2021

'The idiot brain: a neuroscientist explains what your head is really up to', Dean Burnett, 2016

'The Language Instinct', Stephen Pinker, 1994

'The man who mistook his Wife for a hat', Oliver Sacks, 1987

'The State of the Nation: Stroke Statistics February 2018' https://www.stroke.org.uk/system/files/sotn_2018.pdf

'The structure of human abilities', Philip.E.Vernon, 1965

'What it's like to be a dog: and other adventures in animal neuroscience', Gregory Berns, 2017

'Why we sleep: the new science of sleep and dreams', Matthew Walker, 2017

Acknowledgements

Firstly I must thank my Wife, Dee, without whom I would not be here today. At least not with much of my brain intact (I leave it for others to judge whether that is a good thing).

Secondly to my immediate family, whose love and initiative saved me again six months later.

My junior school teacher, Mrs Monk, instilled in me a curiosity and habit of critical thinking which has served me well for over 55 years.

More recently, all of the friends and dedicated professionals who have helped me to be the best that I can.

Special thanks are owed to the NHS – especially the stroke specialist teams at William Harvey Hospital, Ashford, and Eastbourne District General Hospitals, the East Sussex stroke rehabilitation team - and to the Stroke Association.

I have aimed to give due credit for others' work but, if I have forgotten any, or given incorrect attributions, then I apologise, and will be happy to set the record straight in subsequent edition(s).

Milton Keynes UK
Ingram Content Group UK Ltd.
UKHW020117030823
426179UK00006B/173